Design and Optimization of Production Lines

Design and Optimization of Production Lines

Editors

Paolo Renna
Michele Ambrico

MDPI • Basel • Beijing • Wuhan • Barcelona • Belgrade • Manchester • Tokyo • Cluj • Tianjin

Editors

Paolo Renna
School of Engineering,
University of Basilicata
Italy

Michele Ambrico
CRF WCM Research and Innovation,
Campus Manufacturing Melfi
Italy

Editorial Office
MDPI
St. Alban-Anlage 66
4052 Basel, Switzerland

This is a reprint of articles from the Special Issue published online in the open access journal *Applied Sciences* (ISSN 2076-3417) (available at: https://www.mdpi.com/journal/applsci/special_issues/Production_Lines).

For citation purposes, cite each article independently as indicated on the article page online and as indicated below:

LastName, A.A.; LastName, B.B.; LastName, C.C. Article Title. *Journal Name* **Year**, *Volume Number*, Page Range.

ISBN 978-3-03943-961-4 (Hbk)
ISBN 978-3-03943-962-1 (PDF)

Contents

About the Editors

Paolo Renna is an Associate Professor at the School of Engineering of the University of Basilicata. His academic research principally deals with the development of innovative negotiation and production planning in distributed environments. Several of his contributions have been presented on designing multi-agent architecture and testing by discrete event simulation in business-to-business environments. Among these contributions, he is co-author of two research books about e-marketplaces and production planning in production networks. Moreover, he has developed coordination approaches in multi-plant production planning environments and innovative scheduling approaches in flexible and reconfigurable manufacturing systems.

Michele Ambrico is currently a Manufacturing Methodologies Specialist at CRF/WCM Campus Manufacturing in Melfi. He primarily works on numerical simulation for process/product optimization. He received his PhD at the Department of Environmental Engineering and Physics in the Engineering Faculty of the University of Basilicata (Italy), where he also received his degree in Industrial Engineering. His academic studies are focused on the design of robust cellular manufacturing systems and the development of simulation models in dynamic environments. He collaborates with the manufacturing system research group at the School of Engineering (University of Basilicata) on the development of simulation environments for the design and control of manufacturing systems.

Editorial

Special Issue "Design and Optimization of Production Lines"

Paolo Renna [1,*] and Michele Ambrico [2]

[1] School of Engineering, University of Basilicata, 85100 Potenza, Italy
[2] CRF WCM Research and Innovation, Campus Manufacturing Melfi, 85025 Potenza, Italy; michele.ambrico@crf.it
* Correspondence: paolo.renna@unibas.it

Received: 16 November 2020; Accepted: 19 November 2020; Published: 23 November 2020

The classical models for designing production lines follow the objective of balancing the line so as to improve the throughput. Nowadays, the development of sustainable processes is a strategic aspect for the manufacturing industry, and it is a central theme in current innovation projects; the industrial manufacturers study energy-efficient models because of the costs and environmental impact of energy consumption. The latest trends of design and optimization models include the management of reconfigurable machines, switch-off policies, buffer control, and so on, to increase robustness and modularity of production lines and reduce energy consumption.

The "Fourth Industrial Revolution" (alternatively known as "Industry 4.0") supports innovative models for energy consumption and fault tolerance in automated lines, and this drives the changes in design and optimization models for the production lines.

To meet the objective of sustainable production lines in terms of energy consumption, peak electricity demand and energy efficiency including Industry 4.0 technologies, new innovative models are needed to support the design and management of production lines.

This book includes a series of five research studies that reveal new knowledge about the design and management of sustainable production lines.

The topics covered span many diverse areas associated with the design and management of production lines, such as: production improvement in uncertain environments [1], design of production lines to introduce switch-off policies [2], new technologies [3], statistical data analysis [4,5].

In combination, these complementary contributions provide a substantial body of knowledge in the context of production lines that are currently undergoing an epochal industrial transformation.

Funding: This research received no external funding.

Acknowledgments: This publication was only possible with the invaluable contributions from the authors, reviewers, and the editorial team of Applied Sciences.

Conflicts of Interest: The authors declare no conflict of interest.

References

1. Koo, P.-H. A New Self-Balancing Assembly Line Based on Collaborative Ant Behavior. *Appl. Sci.* **2020**, *10*, 6845. [CrossRef]
2. Renna, P.; Materi, S. Design Model of Flow Lines to Include Switch-Off Policies Reducing Energy Consumption. *Appl. Sci.* **2020**, *10*, 1475. [CrossRef]
3. Aleksandrova, I.; Koshelev, E.; Koresheva, E. In-Line Target Production for Laser IFE. *Appl. Sci.* **2020**, *10*, 686. [CrossRef]

4. Zamora-Antuñano, M.A.; Cruz-Salinas, J.; Rodríguez-Reséndiz, J.; González-Gutiérrez, C.A.; Méndez-Lozano, N.; Paredes-García, W.J.; Altamirano-Corro, J.A.; Gaytán-Díaz, J.A. Statistical Analysis and Data Envelopment Analysis to Improve the Efficiency of Manufacturing Process of Electrical Conductors. *Appl. Sci.* **2019**, *9*, 3965. [CrossRef]
5. Murovec, J.; Kušar, J.; Tomaž Berlec, T. Methodology for Searching Representative Elements. *Appl. Sci.* **2019**, *9*, 3482. [CrossRef]

Article

A New Self-Balancing Assembly Line Based on Collaborative Ant Behavior

Pyung-Hoi Koo

Department of Systems Management & Engineering, Pukyong National University, Busan 48513, Korea; phkoo@pknu.ac.kr; Tel.: +82-51-629-6485

Received: 29 August 2020; Accepted: 28 September 2020; Published: 29 September 2020

Abstract: In most mass-production assembly lines, workers perform a set of tasks repetitively predefined by assembly line balancing techniques. The static task assignment often leads to low productivity when the assembly system faces disruptions or uncertainties such as machine breakdown and uneven worker capabilities. The idea of bucket brigades (BB) has been introduced to address the static assignment problems where cooperative behavior of ants is applied to flow line control. This paper examines possible efficiency losses associated with the existing BB-based assembly cell and presents an improved version for assembly cells under uncertain environments. The new system attempts to enhance productivity by assigning assembly tasks to workers dynamically and possibly adding buffers for decoupling consecutive workers. The proposed assembly system is evaluated through simulation experiments under various manufacturing environments. The experimental results show that the new system provides higher productivity than the naïve BB-based assembly cell as well as traditional assembly cells, especially for uncertain assembly environments.

Keywords: assembly systems; assembly cell; line balancing; dynamic task assignment; bucket brigades; self-balancing; autonomous control

---

1. Introduction

A common assembly system consists of m workers (or workstations) arranged along a production line. The components are fed into the line and progressively assembled as they move down from station to station. At a workstation k (= 1, ..., m), a worker performs a set of assembly tasks, S_k, repeatedly. It takes station time, $t(S_k) = \sum_{j \in S_k} t_j$ to complete all the tasks assigned to workstation k where t_j is the assembly time for task j (=1, ..., n). The cycle time, c, of the assembly line is determined by a bottleneck station whose station time is the largest, i.e., $c = max_k\, t(S_k)$. The capacity of the assembly line is $1/c$. When an assembly line is operated in a full capacity, station k with $t(S_k) < c$ has an idle time of $c - t(S_k)$ in each production cycle. Line balance efficiency E is defined by $E = \frac{T}{mc}$ where $T = \sum t_j$. The assembly lines are widely employed in various industries for the production of standardized products.

One of the important decisions in the assembly line involves the work assignment regarding how much work to assign to each worker. Decisions about assembly task assignments are made through assembly line balancing (ALB) techniques whose key objective is to obtain a balanced assembly line by assigning assembly tasks to each worker as evenly as possible. The existing studies on ALB problems are well summarized in survey papers [1–4]. Most existing studies on ALB problems deal with a static line-balancing solution with which the workers perform a pre-assigned set of assembly tasks from cycle to cycle. In general, the ALB decisions are made at the early stage of the assembly line design process in which assembly time for an individual task is assumed to be constant. However, the real industrial environment is often different from the situations on which the ALB decisions are based. Manufacturing systems undergo various types of uncertainties including a machine breakdown, different worker skills, rework requirements, and quality problems. With the uncertainties, the assembly time fluctuates over time so that the assembly lines often fail to achieve the performance as expected [5]. When a new product is introduced, which frequently happens in today's fast-moving market, the assembly systems should be reconfigured. The frequent reconfiguration results in low performance in the traditional assembly line with a static task assignment. Moreover, it is common that the workers have different work speeds and each worker cannot perform assembly tasks at a constant speed throughout a day. When uncertainties are common, it has become more important for production systems to have the flexibility to respond quickly to uncertainties. Assembly cells have been introduced to solve these problems of assembly lines (e.g., [6–9]). A long assembly line is reconfigured into several small assembly cells in each of which a small number of workers (or a single worker in an extreme case) perform the entire range of assembly tasks for a final product. The assembly cells are now widely implemented in industries because they provide better manufacturing performance than traditional assembly lines in terms of productivity, product quality, and customer responsiveness. However, the assembly cell still has a workload balancing problem, especially under system uncertainty.

To solve the balancing problems associated with static task assignments, a new assembly line control method has been introduced where the idea of a bucket brigade (BB) is used as control logic in assembly lines [10–14]. The BB-based assembly cell imitates a cooperative behavior of ants in nature, as shown in Figure 1a. When some ant colonies find a large amount of food in one place, they move the food items to their nest in a collaborative way as follows [15]: An ant carries her food item along the trail. When she encounters a downstream ant (i.e., an ant closer to the nest), she hands over the food item to the downstream ant and returns back toward the food source to move another food item. The recipient ant carries the food item further down the trail to the nest until she encounters an unladen ant (or the nest), and so on. In this control logic, the amount of work that an ant performs depends on her capability and system status (e.g., a road condition). Stronger ants can perform more tasks while weaker ants less.

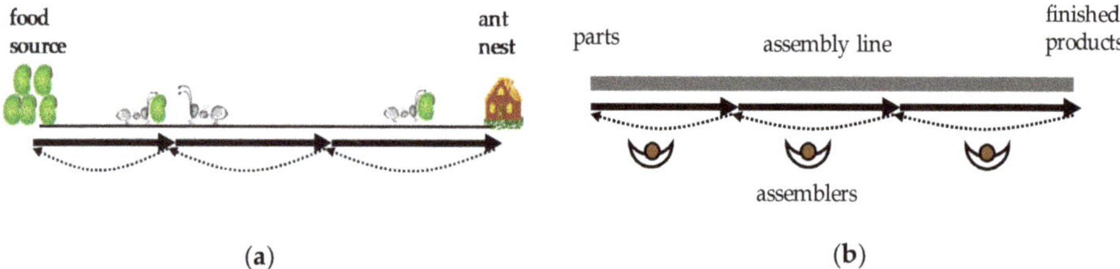

Figure 1. Bucket brigades (BBs): (**a**) cooperative food transport of ants in nature and (**b**) BB-based assembly cell.

Figure 1b shows a BB-based assembly cell. The following is how the BB-based cells work [16]: *"Each worker assembles a product towards completion. Every worker except for the last one hands his current job over to his immediate successor (downstream worker) who is available and goes back upstream to take a new assembly job from his immediate predecessor (upstream worker). The last worker completes assembling all the remaining assembly tasks for a product, puts the final product in the output storage, and walks back upstream. After handing an assembly job over to his immediate successor, the first worker walks back to the beginning of the line to start a new assembly job."* The BB-based assembly cell is characterized as distributed and autonomous control. There is a reduced need for central planning and management because bucket brigades make the flow line self-balancing. The workers do not perform only the jobs previously assigned to them but perform the jobs that may be different from cycle to cycle. The balancing problems are mitigated by the self-balancing property. Each worker performs an assembly task with a velocity commensurate with his own capability. The throughput may be increased because bucket brigades spontaneously generate the optimal division of work. However, in real-world assembly operations, some situations undermine the performance of the BB-based cell. Two kinds of efficiency losses have been identified when the concept of the BB is applied to industrial systems: task hand-offs and worker blocking [17]. Let us consider a BB-based assembly line in which each assembly task should be performed by only one operator without any interruption (i.e., the task is not divisible.) In this case, once a worker starts an assembly task on a product, he should complete the task. Then, it may happen that a downstream worker cannot take over an assembly task from the upstream worker immediately after he is available. Then, the downstream worker should wait (hand-off loss) until the predecessor finishes his current job. Moreover, the assembly time for a task is often not constant but variable over time. In this case, a situation may happen that an upstream worker cannot start a new assembly task (blocking loss) when his downstream worker is struggling to finish the same task.

This paper examines performance losses including hand-offs and blocking in the existing BB-based assembly cell and presents an improved version of the BB-based cell to reduce performance losses. The proposed system aims to enhance productivity by assigning assembly tasks to workers dynamically and possibly adding intermediate buffers for decoupling consecutive workers. The assembly systems under consideration have the following assumptions:

- A single type of product is assembled in the assembly system.
- Multiple elementary assembly tasks are needed to produce a final product.
- Each task is indivisible in assembly. A half-finished assembly task cannot be transferred to the downstream worker.
- Assembly time for a task may not be deterministic but stochastic with some variabilities.
- The sequence of assembly tasks is predefined on the basis of their precedence constraints.
- The sequence of the worker is fixed (no upstream worker can pass through his downstream worker).
- Two workers are not allowed to perform the same assembly task at the same time.

This paper is organized as follows. Section 2 reviews the literature on previous studies on BB-based assembly systems. In Section 3, we present the problems associated with efficiency losses in BB-based cells. A production capacity model in BB-based cells is introduced by considering hand-off and blocking delays. Section 4 presents an improved version of the existing BB-based cell, and Section 5 compares the performance of the new assembly cell with that of the other assembly systems through simulation experiments. Finally, we give concluding remarks in Section 6.

2. Literature Review

Assembly line balancing (ALB) is a practical industrial problem that should be solved in many different industries. The main task of ALB decisions is to allocate elementary assembly tasks to workers (or workstations), typically intending to maximize line efficiency. As there are many different assembly environments, the research works on the ALB problems have extensions including multi-manned workstations [18,19], multi-product assembly lines [20,21], U-shaped assembly lines [22,23], asynchronous assembly lines [24], and robotic assembly lines [25]. Most existing studies deal with ALB problems by using optimization and/or heuristic models. The models provide static line-balancing solutions with which the workers perform a pre-assigned set of assembly tasks from cycle to cycle. Even though the optimization/heuristic models are expected to perform well under static manufacturing environments, high performance may not be realized under dynamic manufacturing environments.

A BB-based assembly cell has been introduced where assembly tasks are not assigned to a worker in a static way but can be performed by any worker [16]. Each worker follows a simple control logic, which is discussed in Figure 1b. The authors of [16] claim that the BB-based system can achieve the highest productivity when the following three conditions are satisfied: (1) the walk-back time is ignorable, (2) the task can be handed over immediately when the downstream worker is available, and (3) each assembler $i = 1, \dots, m$ is characterized by a distinct constant work velocity v_i. The work velocity for each assembler is deterministic. It is proved under the three ideal conditions that if the workers are sequenced along the flow line from slowest to fastest, the system becomes self-balanced such that the hand-offs between any two consecutive workers take place at a fixed location, and the production rate per unit time converges to $\sum_{i=1}^{n} v_i$, which is the maximum attainable from the given workers [16]. The problem is that the ideal conditions are not satisfied in most real-world production environments.

Reference [26] shows that the BB-based system can be effective even in the presence of stochastic task times. Reference [27] presents a case study where a traditional tractor assembly line was replaced by a BB-based assembly line. Here, walk-back times were explicitly considered. It is reported that the productivity increased by adopting the BB concept more than compensated for the time lost by walk-back times. Reference [28] examined the effect of labor turnover on the performance of different assembly methods. A simulation study showed that the BB-based assembly line outperformed a traditional assembly line with fixed work allocation especially when the labor turnover rate was high. Reference [10] explores how the BB protocol can be extended to make subassembly lines self-balancing where all the subassembly lines should be synchronized to produce at the same rate. Reference [29] examined a two-worker BB system where one worker was faster than the other over some part of the production line and slower over another part of the line. Reference [30] addresses blocking problems in a BB-based production system where blocking occurred only when the preceding worker was faster than the downstream worker. Deterministic task time was assumed. Initial positions of workers were identified where no blockings occurred. Reference [31] introduces a case study where a BB-based system was applied to an assembly line of fashion bags, showing the advantages of a BB-based system in terms of flexibility, low work-in-process (WIP) levels, and the ability to handle uncertainties. In the warehouse order picking environment, analytical models have been presented to quantify the losses by hand-off delay [32] and blocking delay [33]. References [14,34] address the walk-back time problems in a BB-based production system. A new design of bucket brigade was proposed to reduce unproductive travel with a U-shaped production cell where workers work on one side of the aisle for the forward direction and they work on the other side for the reverse direction. Reference [14] explicitly considers hand-off time and walk-back time in a U-shaped BB system. It is argued that the U-shaped BB system can be more productive than the traditional BB system if the number of workers in the cell is small. The uncertainties in the manufacturing system are not considered in the model.

As reviewed above, many studies have focused on analytical models dealing with the behavior of the BB-based system where each worker has a distinct and constant work speed, and the workers are sequenced from slowest to fastest. Although some research works address practical issues by presenting improved versions of BB systems by applying BB protocols to U-shaped assembly cells [14] and sub-assembly cells [10], little work has been done for addressing hand-off and blocking problems in existing BB-based systems under uncertain environments. This paper aims to fill the research gap of the existing works. We examine possible efficiency losses associated with hand-off and blocking delays in the existing BB-based assembly system and present a production capacity model in BB-based cells with hand-off and blocking delays. An improved BB-based system is introduced for assembly cells under an uncertain environment that attempts to enhance productivity by assigning assembly tasks to workers dynamically and possibly adding buffers for decoupling consecutive workers. We also examine the effect of the sequence of workers in the new BB system.

3. The Efficiency of BB-Based Assembly Cell

We now examine efficiency losses in the BB-based system caused by hand-offs and worker blocking in this section on assembly lines under uncertainties. Let us first examine the effect of the uncertain events on the assembly time distribution. The uncertain events in assembly systems include assembly tool breakdown, defected parts, workers' job skills, and utility problems, to name a few. Even though the net assembly task time is deterministic, the uncertain events make effective assembly times stochastic. Suppose net assembly time for a task is deterministic with t_s and an assembly tool is broken in the middle of an assembly activity. Then, the assembly time is expanded to the net assembly time plus an independent tool replacement time. Suppose the tool breakage takes place with probability p and it requires t_r minutes on average to fix the problem. The effective task time will be t_s with probability 1-p and $t_s + t_r$ with probability p. On average, the effective assembly time, τ_s, will be $\tau_s = (1 - p)t_s + p(t_s + t_r) = t_s + p{\cdot}t_r$. Now, suppose the tool breakage time is stochastic with a coefficient of variation, v_r. The coefficient of variation (CV) is defined as the standard deviation over the average of a variable. Then, the CV of the effective assembly time, v_s, is calculated by using the model presented in [35] as follows:

$$v_s = \sqrt{\frac{t_s^2 + p_s\left(v_r^2\, t_r^2 + 2t_r t_s + t_r^2\right)}{\tau_s^2} - 1} \quad (1)$$

To understand the effect of uncertain events on the effective assembly time, let us take a simple example. Suppose the net assembly time is constant with 3.0 min, and it undergoes a tool breakdown with a probability of 0.02 following an exponential distribution of average 9.0 min. Then, the average assembly time is prolonged from 3.0 min to 3.18 min (= 3.0 + 0.02 · 9.0) and the CV of the effective assembly time is 0.56, calculated from Equation (1). Hence, even with a constant net assembly time, it is reasonable to assume that the effective assembly time is not deterministic but stochastic in the system under uncertainties.

Figure 2 shows a situation where a hand-off loss happens. Suppose the upstream worker is performing an assembly task starting at time *ta*, whose expected assembly time is (*tc* – *ta*) time units. If the downstream worker becomes available and ready to take over a new assembly job at time *tb* from the upstream worker who is performing the assembly task, he should wait until the upstream worker finishes his current task. Since the current task is expected to be finished at time *tc*, the expected waiting time of the downstream worker for the hand-off is (*tc* – *tb*) time units.

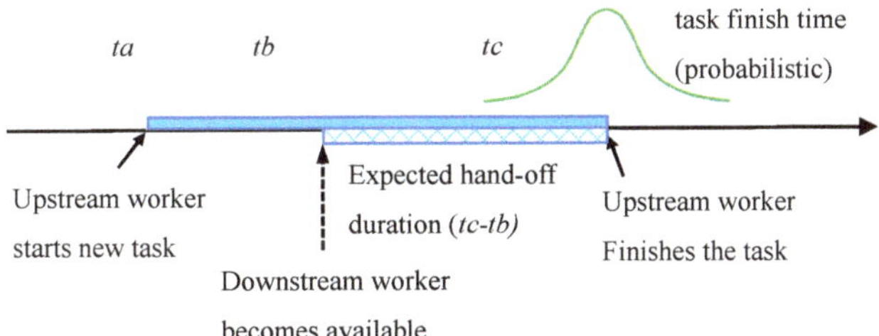

Figure 2. Hand-off loss in BB-based assembly cell. The expected hand-off time is *tc* – *tb*.

We now estimate the performance of workers when the hand-off times are present. We will adopt the result of [36] where it is proved that the expected time per each hand-off, h_o, is $h_o = \frac{E[t^2]}{2E[t]}$, where t is a variable representing assembly time. Since $E[t^2] = V(t) + E[t]^2 = (cv(t)^2 + 1)\, E[t]^2$ and $E[t] = T/n$, we have

$$h_o = \frac{E\left[t^2\right]}{2E[t]} = \frac{\left(cv(t)^2 + 1\right)T}{2n} \quad (2)$$

where $V(t)$ is the variance of t, $cv(t)$ is the CV of t, and T is the total assembly time for a final product. It is seen from Equation (2) that there is a hand-off time even with no assembly time variability with $cv(t) = 0$.

Consider an m-worker assembly line dedicated to a product type where the final product requires n assembly tasks. When all the workers are fully utilized, then the maximum production rate per unit time of the assembly line is m/T. In the BB-based cell, each product experiences $(m-1)$ hand-offs in the m-worker line. Then, the total expected hand-off time experienced by a final product, h, is as follows:

$$h = (m-1)\frac{E[t^2]}{2E[t]} = \frac{(m-1)\left(cv(t)^2+1\right)T}{2n} \quad (3)$$

To complete assembling a product on a BB-based cell, assemblers spend T time units for performing assembly tasks and h time units being idle. Hence, $(T + h)$ time units are required to complete a product. Out of $(T + h)$ time units, only T time units are value-added time. The efficiency loss percentage due to hand-off delays, E_{loss}, is $E_{loss} = \frac{h}{T+h}$. Let the average efficiency of workers be E_o. Then, E_o can be obtained by $E_o = 1 - E_{loss} = \frac{T}{T+h}$. Then, the production capacity of m-worker BB-based cell, r, is as follows:

$$r = E_0\frac{m}{T} = \frac{m}{T+(m-1)\left(cv(t)^2+1\right)T/(2n)} = \frac{m}{T\left[1+(m-1)\left(cv(t)^2+1\right)/(2n)\right]} \quad (4)$$

From Equations (3) and (4), it is seen that the production capacity of the BB-based cell is a function of the number of assembly tasks (n), the number of workers (m), and the CVs of task times. The production capacity increases as the number of assembly tasks increases and the number of operators decreases. Figure 3 shows the effects of the number of workers and the number of assembly tasks on the efficiency loss caused by hand-offs in the case of (a) deterministic task time with $cv = 0.0$ and (b) stochastic task time with $cv = 1.0$. When the number of assembly tasks is large and the number of operators is small, the proportion of the hand-off time in the total assembly time becomes small, thus decreasing the productivity loss. Note that n/m indicates the average number of assembly tasks handled by a single worker. If n/m becomes very large, the efficiency loss caused by hand-offs approaches zero. The results show that assigning a small number of operators to the BB-based cell is preferable. Comparing Figure 3a,b, we can also find that the system with higher variability of task time leads to higher hand-off losses.

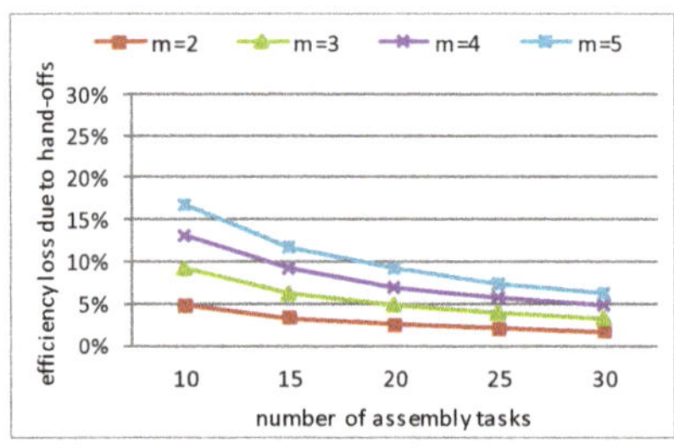

(**a**) deterministic task time (cv = 0.0)

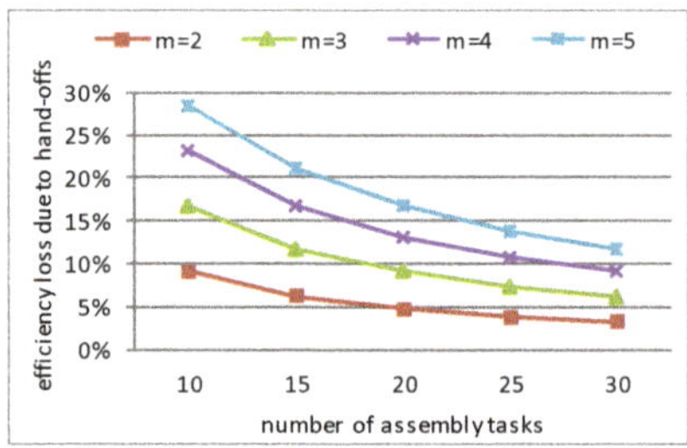

(**b**) stochastic task time (cv = 1.0)

Figure 3. Efficiency loss in BB assembly cell with hand-off times under (**a**) deterministic task time (cv = 0.0) and (**b**) stochastic task time (cv = 1.0). It is seen that the hand-off losses are increased by (1) higher task time variabilities, (2) more workers in a cell, and (3) fewer tasks assigned to a worker.

System efficiency can be also undermined by blocking phenomena in the BB-based cell. When an upstream worker finishes an assembly task but finds his downstream worker performing the next task, he cannot start the next assembly task but wait until the downstream worker has finished his current job. Even though each product requires the same assembly tasks, the blocking may happen when assembly times are not deterministic. (Note that a major reason for blocking in a warehouse order picking system is different order contents form order to order [33]). The blocking not only has an effect on production capacity but also production lead time. Simulation experiments are performed to examine the blocking time in a three-worker BB-based cell where a single product type requiring 12 assembly tasks is assembled. Figure 4a shows the ratio of blocking time over the total net assembly time of a product over different assembly time variabilities. It is found in Figure 4a that when the assembly times are deterministic with cv = 0.0, no blocking occurs. As the assembly time variabilities increase, blocking time increases (Figure 4a), resulting in a lower production rate (Figure 4b).

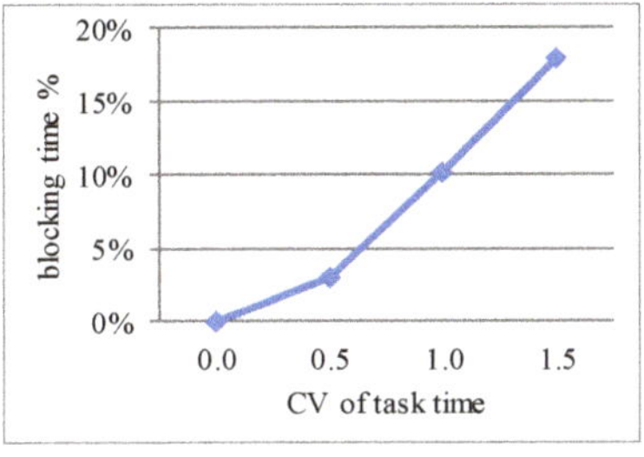

(a) blocking time percentage

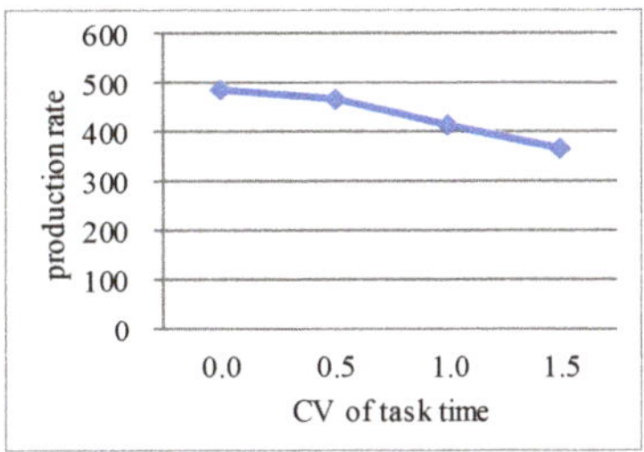

(b) production rate

Figure 4. Effect of blocking over different assembly time variabilities in terms of (**a**) blocking time percentage over total assembly time and (**b**) production rate.

4. A New Control Method of BB-Based Assembly Cell

We examined efficiency losses caused by hand-offs and blockings in the BB-based cells in the previous section. To handle efficiency problems in the existing BB-based assembly cell, this paper presents an improved version. (For comparison, the existing naïve BB-based assembly cell and the newly proposed BB assembly cell are called NBB and ABB cell, respectively). The idea for the new model was obtained from the BB-based system in [17], where a BB-based order picking and a zone picking were integrated to reduce the hand-off and blocking losses in the order picking system. Figure 5 illustrates a three-worker ABB cell to assemble a product type with 12 assembly tasks. The major difference between NBB cell and ABB cell is twofold: (1) there is a limited set of tasks that each worker can perform in ABB cell, and (2) there may be buffer space at the end of each worker except the last worker. Each worker is not allowed to perform assembly tasks out of the assigned set. In Figure 5, worker 1 is allowed to perform assembly jobs between task A1 and A6, worker 2 between A2 and A9, and worker 3 between A3 and A12.

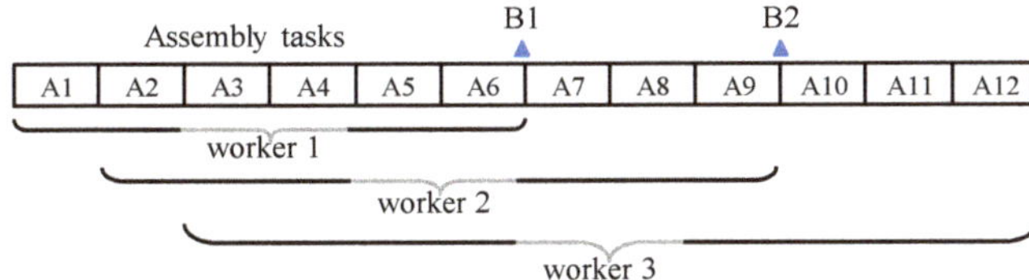

Figure 5. Task assignment of workers in the proposed BB assembly (ABB) cell. B1 and B2 indicate the buffers.

In an ordinary situation, the workers follow a naive control logic of the NBB cell. The new control logic comes into effect when a worker finishes all the tasks within the assigned tasks but his immediate downstream worker is not available. For example, if worker 1 (W1) finishes all the tasks up to task A6, but finds worker 2 (W2) busy, then W1 cannot start performing a new assembly task beyond A6 but waits until W2 is available to take his job. (Note that in the NBB cells, W1 will attempt to start A7 assembly tasks.) When W2 is available, W1 hands over his semi-finished assembly to W2 and returns back upstream to the beginning of the assembly line to start the first assembly task for a new product. If buffer space is allowed at buffer B1, then W1 puts the semi-finished assembly at buffer B1 after finishing all the tasks assigned to him and returns to the beginning of the assembly line. When ready for the next assembly job, W2 moves upstream and checks if there is any semi-finished product in buffer B1. If there exists at least one workpiece at B1, W2 starts the assembly task with it. Otherwise, as in the NBB cell, W2 goes upstream to take over the assembly job from W1. Worker 3 (W3) completes assembly jobs by performing task A12 and puts the finished product at output storage and moves upstream. If there is a semi-finished assembly at buffer B2, W3 starts the assembly task with it. Otherwise, W3 goes upstream to take a job from W2. Note that if there is a workpiece in a buffer, it means that the upstream worker does not experience blocking and the downstream worker does not experience a hand-off time loss. Hence, efficiency losses caused by blocking as well as hand-offs are expected to be reduced with the new system. With each worker assigned to limited work contents, they perform assembly tasks far from each other, possibly resulting in less blocking frequencies.

5. Simulation Experiments and Results

Simulation studies were conducted using a simulation package, ARENA, to evaluate the performance of the proposed assembly method. The experiments were performed with assembly systems where only one product type (requiring 12 assembly tasks) was assembled. The performance of the proposed BB-based (ABB) assembly cell was compared with those of the general assembly cell (GEN cell) and NBB assembly cell. Three workers were sequenced along the assembly line to perform assembly tasks in a cell. The term "cell" is used because a small number of workers are in charge of an assembly line. In the GEN cell, each worker performed four pre-assigned assembly tasks. Ten different simulation scenarios were created with assembly task times randomly generated between 3 and 5 minutes. The dataset is given in Table 1. It is assumed that the work sequence of each task was predefined by the precedence constraints of assembly tasks. For stochastic assembly times, a Gamma distribution was assumed because the Gamma distribution is widely used in assembly line analysis and easy to define assembly time variabilities. Simulation experiments were carried out 10 times for each scenario in order to have meaningful statistics. Each simulation run covered 30 days. To collect steady-state data, this study discarded statistics for the first 10 days (4800 min), and data for the next 20 days (1 month, 9600 min) were collected for analysis.

Table 1. Assembly operation data for 10 different scenarios.

Task Number	Scenarios									
	1	2	3	4	5	6	7	8	9	10
1	4.0	4.8	3.8	3.3	4.4	4.5	3.2	4.7	5.0	4.6
2	3.3	3.2	4.0	5.0	4.0	3.6	4.9	4.4	4.2	4.8
3	4.0	3.1	5.0	3.2	3.7	3.3	4.1	4.1	3.3	3.3
4	3.8	4.8	5.0	3.6	4.0	4.1	3.7	4.5	3.7	5.0
5	3.9	4.0	4.4	3.2	3.3	4.3	3.0	3.8	4.3	4.9
6	4.5	4.7	3.1	4.1	3.6	4.3	3.0	3.1	4.4	4.0
7	5.0	3.5	3.3	4.5	5.0	3.2	4.1	3.9	4.8	3.8
8	4.5	4.9	4.9	3.6	4.3	4.0	3.1	3.1	3.6	3.4
9	3.2	3.2	3.4	4.9	3.6	3.9	3.5	3.6	4.9	4.9
10	3.8	3.3	4.4	3.1	4.3	4.9	3.2	3.2	4.4	4.2
11	4.1	4.9	4.2	4.8	5.0	4.5	4.6	3.1	4.4	3.0
12	3.8	4.6	4.2	5.0	3.7	3.8	4.7	4.6	4.8	3.2 [1]

[1] The values in the cells indicate the expected assembly time in minutes for each task.

5.1. Experimental Analysis with Deterministic Task Times (No Uncertainty)

When the assembly task time was deterministic, the performance of the GEN cell could be obtained analytically. To understand the performance of the different assembly strategies clearly, this study takes scenario 1 as an example, whose total assembly time was $T = 47.9$ min.

GEN cell: In GEN cell, the task times of an assembly cycle for workers 1, 2, and 3 were 15.1, 17.9, and 14.9 min, respectively. Hence, the bottleneck resource in the GEN cell was worker 2. The system efficiency of the GEN cell was 0.892 (= 47.9/(3 × 17.9)) and the production rate of the three-worker GEN cell was 536/month (= 9600 min/17.9 min).

BB-based cell (NBB cell, ABB cell): From Equation (3), the hand-off time for each assembly cycle was $h = E(t) = 3.99$. Then, the line efficiency was $E_0 = T/(T + h) = 47.9/(47.9 + 3.99) = 0.923$. Note that with deterministic task times, blocking is not likely to happen. From Equation (4), the production rate of the three-worker BB-based cell was 0.0578 units/min, which meant 555 units/month.

It should be noted that in the deterministic task time environment, the efficiency losses in the GEN cell are due to line imbalance while the efficiency losses in the NBB and ABB cells are caused by hand-offs. Figure 6 compares the performance of different assembly methods. When workers are utilized 100%, the production capacity will be 601 units/month (= 9600 × 3/47.9). It is seen that the balancing losses existed in the GEN cell while hand-off losses were present in the NBB cell and ABB cell.

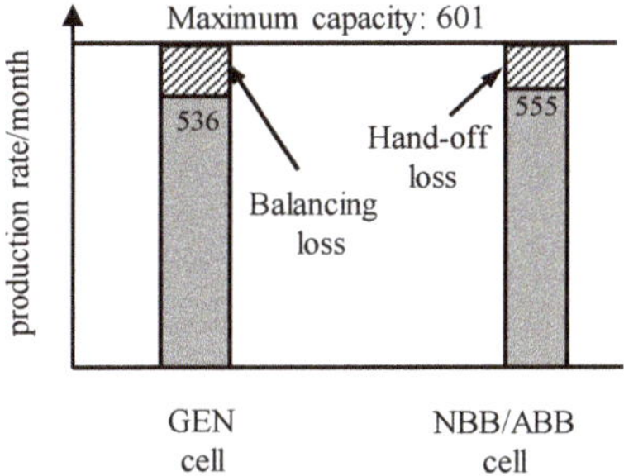

Figure 6. Performance of assembly methods under deterministic task time.

5.2. Experimental Analysis with Stochastic Task Times

As discussed previously, in a real-world manufacturing environment, it is common for the effective assembly times to be stochastic. Reference [37] classified variability levels in manufacturing systems into low, moderate, and high, on the basis of the magnitude of the CV values: low variation for CVs less than 0.75, moderate variation for CVs between 0.75 and 1.33, and HV high variation for CVs greater than 1.33. In this study, the performance of different assembly systems was examined with four *CV* values: 0.0, 0.5, 1.0, and 1.5. It was assumed in the ABB assembly cell that the assembly tasks assigned to the workers were the same as in Figure 5 and that no intermediate buffers existed. (The effect of in-between buffers will be discussed in the next subsection).

Figure 7 compares the performance of the different assembly systems. Assembly jobs were transferred to downstream operators when all the workers were ready to start new assembly jobs in the GEN cell or when the immediate downstream worker was ready to take a new job in the NBB cell and ABB cell. When there was no variability (i.e., deterministic task time with CV = 0), not much performance difference could be seen among the three cell control methods. As the variabilities of the task times increased, the performance became worse in every assembly system. The higher variabilities of task times aggravated workload imbalance in the GEN cell because the line balancing efficiency changed due to stochastic task times different from fixed standard time assumed in the line design stage. The variability of task times also affected the performance of the NBB cell and the ABB cell because the hand-off and blocking delays were increased with higher task time variabilities (see Equation (3)).

The figure shows that the effect of assembly time variabilities on the performance was less in the NBB cell and the ABB cell compared to the GEN cell. Among NBB and ABB cells, the ABB cell provided consistently better performance than the NBB cell under a stochastic task time environment.

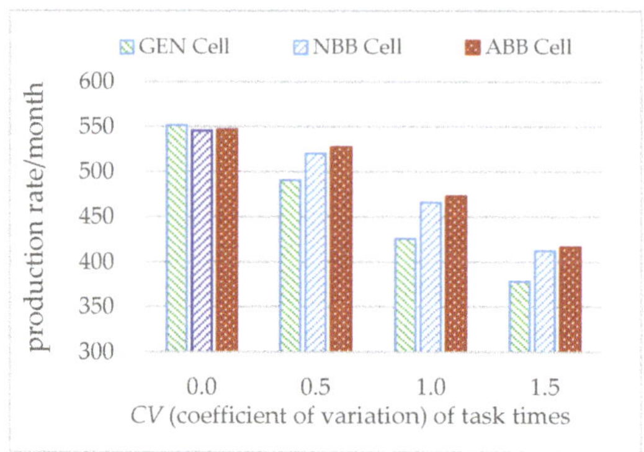

Figure 7. Performance of assembly systems with various assembly time variabilities.

The better performance of the ABB cell over NBB cell was due to reduced hand-off times and blocking times expected in the ABB cell. Figure 8 shows the hand-off times and blocking times in the NBB and ABB cells. It is seen that hand-off times and blocking times were consistently less in the ABB cell than in the NBB cell, and the difference became larger with higher CV values.

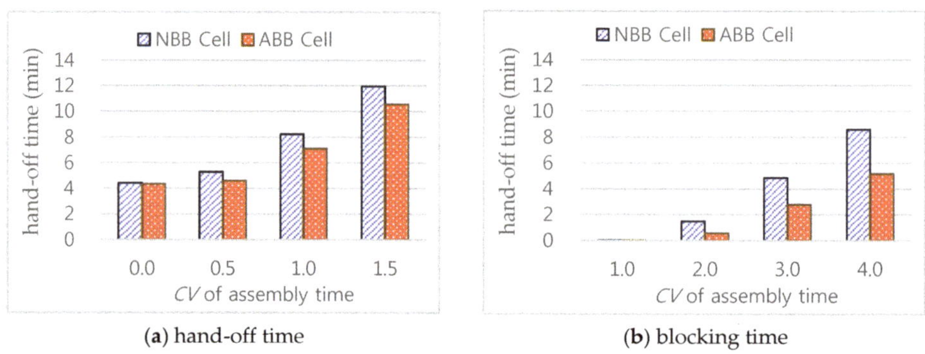

(**a**) hand-off time (**b**) blocking time

Figure 8. Comparison of the performance of naïve BB-based assembly (NBB) cell and ABB cell in terms of (**a**) hand-off time and (**b**) blocking time.

5.3. Performance Comparison of Assembly Systems with Intermediate Buffers

In the previous experiments, this study assumed that there was no in-between buffer. Next, this study examined a new strategy where there were in-between buffers at the end of each work area for workers for the GEN cell and ABB cell. Note that a worker was not allowed to perform assembly tasks out of the assigned set in the ABB cell. When a worker finished the last task of the assigned tasks and found no downstream worker available, he put the half-finished assembly at the in-between buffer (if there was an empty space left) located at the end of his work area and went back upstream to take a new job from his predecessor (or went back to the beginning of the assembly cell for a new product if he was the first worker of the line). If there was no empty space in the buffer, the worker waited until an empty space was available. It was expected that the in-between buffers mitigated some adverse effects of system variabilities. However, if the buffers were too large, it would increase the production lead time and WIP levels. Therefore, it was necessary to maintain an appropriate buffer size.

Figure 9 compares the production rate and lead time over different buffer sizes. Here, the CVs of task times were assumed to be 1.0. It is seen that as the buffer size increased, the production rate increased in all assembly systems except for the NBB cell where a buffer was inherently not allowed. It is seen that the ABB cell performed better than the other systems in terms of the production rate when the buffer size was zero or one. It is seen that only a small buffer size (e.g., one) led to a large increase in throughput rate, with little increase in production lead time in the ABB cell. When the buffer size was larger than one, the GEN cell provided a good performance in terms of production rate with the cost of long lead time. Observing Figure 9b, it is clear that the lead time increased sharply in the GEN cell while the lead time remained almost unchanged in the ABB cell, even with a larger buffer size. The experimental results indicate that the proposed ABB cell is a promising assembly system when the lead time, as well as production rate, is an important performance measure and there is very limited space for intermediate buffers.

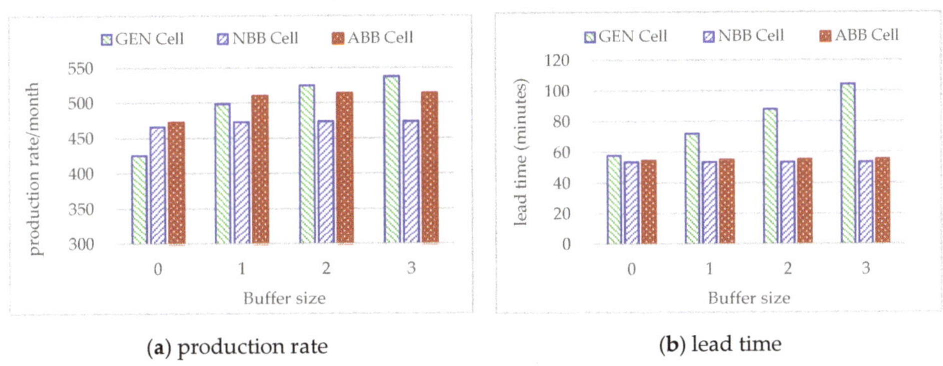

(a) production rate (b) lead time

Figure 9. Performance of different assembly systems over varying buffer sizes in terms of (**a**) production rate and (**b**) production lead time.

5.4. Performance Analysis for the System with Different Workers' Speeds

Most ALB solutions assume that every worker has an identical work speed. However, in real-world manufacturing systems, experienced and unexperienced workers with different work speeds are deployed together in an assembly line. In this case, the production capacity is often restricted by inexperienced and slow workers, resulting in lower productivity than expected. In the BB-based assembly cells, the workers are not assigned a pre-defined set of tasks but perform assembly tasks as much as their ability. The dynamic task assignment is expected to lead to a minimum productivity loss due to different work speeds. This section examines the effect of the different work speeds on the performance of each assembly system. Figure 10 shows the performance of the GEN, NBB, and ABB cells. It was assumed that the assembly tasks are performed by three workers with different work speeds with (a) no buffer and (b) buffer size of one, with an assembly time CV of 1.0. In the figure, speed notation (0.9, 1.0, 1.1) indicates that the work speeds were 0.9, 1.0, and 1.1 for the first, second, and third workers, respectively. Since the buffer was not allowed in the NBB cell, there was no experimental result for the NBB cell in Figure 10b. In every case, the NBB and ABB cells performed better than GEN cells whether the buffer was allowed or not. It is seen that the ABB cell worked better than the NBB cell even when the buffer was not allowed. When buffers were allowed, the ABB cell consistently worked better than the GEN cell. The better performance was realized in the ABB cell no matter where the slower or faster workers were located. The figure shows that the performance of the GEN cell worked well when all the workers' speeds were the same. On the other hand, it is seen that the ABB model worked well when the workers had different work speeds.

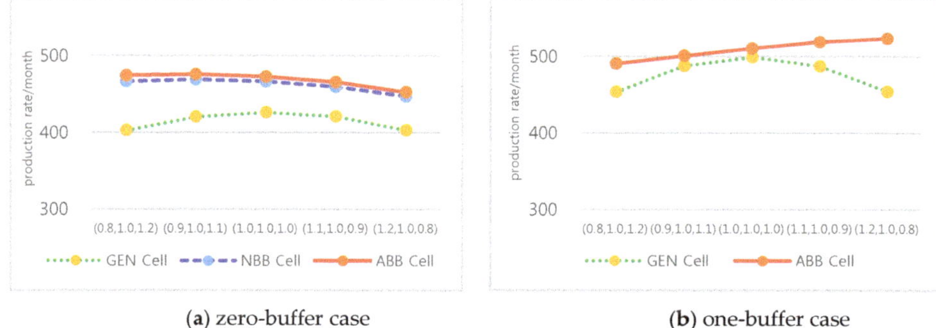

(**a**) zero-buffer case (**b**) one-buffer case

Figure 10. Performance of assembly systems over different work speeds with (**a**) buffer size of zero and (**b**) buffer size of one. The notation at the x-axis, ($s1$, $s2$, $s3$) indicates that the work speeds are $s1$, $s2$, and $s3$ for the first, second, and third workers, respectively.

Figure 10a shows that the ABB cell worked better when faster workers were located at the beginning of the line in the zero-buffer case. This result is contradictory to the claim in [16] that BB-based control works best when workers are sequenced from slowest to fastest along the line. The contradictory result can be explained as follows: since the ABB cell restricts the work contents for a worker to perform, neighboring workers are less likely to perform consecutive assembly tasks that may result in blocking. Hence, blocking delays can be reduced in the ABB cell. The hand-off times depend on the task time of the upstream workers. When the work speed of an upstream worker is fast, then actual assembly time is reduced, resulting in lower hand-off time. Figure 11 shows hand-off and blocking times per product in the ABB cell with (a) zero-buffer and (b) buffer size of one. It is seen that in the ABB cell, hand-off times were much more than blocking times. It is also seen that the hand-off times could be reduced when faster workers were located at the beginning.

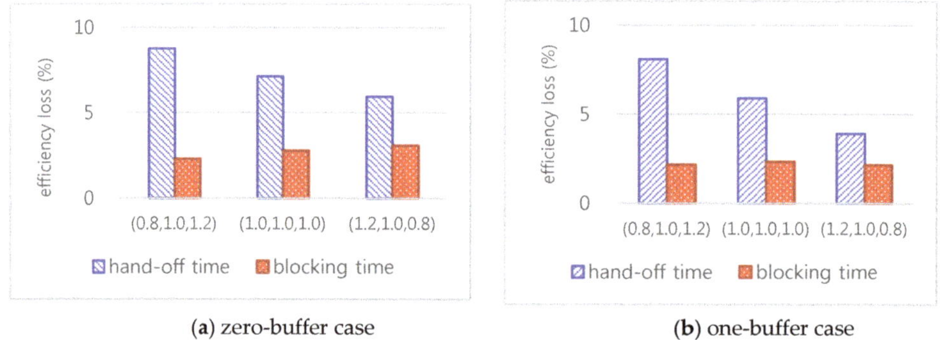

(**a**) zero-buffer case (**b**) one-buffer case

Figure 11. Efficiency losses caused by hand-offs and worker blocking with (**a**) buffer size of zero and (**b**) buffer size of one.

6. Discussion and Conclusions

Once an assembly line is configured on the basis of the assembly line balancing (ALB) decisions, the system is operated in a static way where workers perform the pre-determined set of assembly tasks repetitively. The static decision often leads to low performance in dynamic and uncertain assembly environments. The idea of bucket brigades (BBs) has been applied to assembly systems to respond to system uncertainties. However, the naïve BB-based assembly cell (NBB cell) suffers from some efficiency losses including hand-offs and blocking in most assembly flow lines. This paper presented an improved version (ABB cell) of the NBB cell. The performance of the new assembly cell is compared with the other assembly systems under various assembly environments. From the numerical and experimental analysis, we have produced some interesting findings of the new BB-based assembly system (ABB cell) as follows:

- The key finding of our study is that the ABB cell is robust in that it performs especially well under dynamic and uncertain assembly environments. As the variabilities increase, both the NBB and ABB cells provide worse performance. However, the performance of the ABB cells is less affected by the high variabilities than that of the NBB cells, with fewer hand-off and blocking delays.
- The BB-based cell works well when n/m is large, where n is the number of assembly tasks and m is the number of workers in an assembly cell. As given in Equation (1), the hand-off losses are proportional to the n/m ratio. Parallel to this finding is that assigning a small number of workers to a BB-based cell is preferable.
- Even a minimal buffer size, e.g., buffer size of one, in the ABB cell leads to a big performance increase in terms of production rate. Moreover, the buffer size has little effect on the production lead time in the ABB cells, unlike traditional assembly cells. Hence, the ABB cell can be a good assembly system when the lead time as well as the production rate is a key performance measure.
- Existing research works about the BB-based system assume that the workers are sequenced from slowest to fastest to have a maximum capacity. Our results are contradictory to this assumption. In our BB-based system, the workers may be sequenced in the other way, i.e., from fastest to slowest. It is found that the reverse sequence can reduce hand-off delays and improve performance in some cases (see Figure 10). This finding is important because we can have the flexibility to deploy the workers along the line in the new assembly system.

Although the new assembly system has some advantages over traditional assembly cells and naïve BB-based cells, there are some considerations that should be taken into for the new assembly system to be successfully implemented. One of the important characteristics in the new assembly system involves a distributed and autonomous decision making instead of centralized decision making. Each worker in the production line performs the tasks possibly different from cycle to cycle by following a control logic without any predefined tasks assigned in advance. Hence, workers should be multi-functional and willing to work in collaboration with their colleagues. An appropriate training program for workers should be provided. Moreover, as argued in [38,39], both the benefits and negative aspects of the dynamic work assignments should be taken into consideration together in selecting appropriate assignment strategies. Workers in the BB-based system perform their jobs commensurate with their capabilities. Skilled workers work more while un-skilled workers less. To exert a full capacity of each worker, an appropriate incentive system may be needed.

This study considers assembly systems producing a single product type. It is believed that our model can be extended for multi-product assembly systems with minimum adjustment of the control logic. Walk-back time is ignored in this paper. Hence, the proposed system may be suitable for assembly systems where light and small components are assembled in a short flow line. For the assembly systems dealing with big and heavy components, some new ideas may be needed, which is a possible future work.

Author Contributions: Methodology, Formal Analysis, Investigation, Data Curation, Writing-Original Draft Preparation, Writing-Review & Editing, P.-H.K. The author have read and agreed to the published version of the manuscrip.

Funding: This work was supported by the National Research Foundation of Korea (NRF) grant funded by the Korea government (MSIT) (no. 2019R1F1A1057585).

Conflicts of Interest: The author declares no conflict of interest.

References

1. Hu, S.J.; Ko, J.; Weyand, J.; El Maraghy, H.A.; Lien, T.K.; Koren, Y.; Bley, H.; Chryssolouris, G.; Nasr, N.; Shpitalni, M. Assembly system design and operations for product variety. *CIRP Ann.* **2011**, *60*, 715–733. [CrossRef]
2. Battaia, O.; Dolgui, A. A taxonomy of line balancing problems and their solution approaches. *Int. J. Prod. Econ.* **2013**, *142*, 259–277. [CrossRef]
3. Aguilar, H.; García-Villoria, A.; Pastor, R. A survey of the parallel assembly lines balancing problem. *Comput. Oper. Res.* **2020**, *124*, 105061. [CrossRef]
4. Eghtesadifard, M.; Khalifehb, M.; Khorram, M. A systematic review of research themes and hot topics in assembly line balancing through the web of science within 1990–2017. *Comput. Ind. Eng.* **2020**, *139*, 106182. [CrossRef]
5. Pereira, J. The robust (minmax regret) assembly line worker assignment and balancing problem. *Comput. Oper. Res.* **2018**, *93*, 27–40. [CrossRef]
6. Johnson, D.J. Converting assembly lines to assembly cells at sheet metal products: Insights on performance improvements. *Int. J. Prod. Res.* **2005**, *43*, 1483–1509. [CrossRef]
7. Ruiz-Torres, A.J.; Mahmood, F. Impact of worker and shop flexibility on assembly cells. *Int. J. Prod. Res.* **2007**, *45*, 1369–1388. [CrossRef]
8. Yu, Y.; Tang, J.; Sun, W.; Yin, Y.; Kaku, I. Reducing worker(s) by converting assembly line into a pure cell system. *Int. J. Prod. Econ.* **2013**, *145*, 799–806. [CrossRef]
9. Yu, Y.; Wang, J.; Ma, K.; Sun, W. Seru system balancing: Definition, formulation, and exact solution. *Comput. Ind. Eng.* **2018**, *122*, 318–325. [CrossRef]
10. Bartholdi, J.J.; Eisenstein, D.D.; Lim, Y.F. Bucket brigades on in-tree assembly networks. *Eur. J. Oper. Res.* **2006**, *168*, 870–879. [CrossRef]
11. Gurumoorthy, K.S.; Banerjee, A.; Paul, A. Dynamics of 2-worker bucket brigade assembly line with blocking and instantaneous walk-back. *Oper. Res. Lett.* **2009**, *37*, 159–162. [CrossRef]
12. Koo, P.H. Self-balancing assembly line using bucket brigades. In Proceedings of the 37th International Conference on Computers and Industrial Engineering, Alexandria, Egypt, 20–23 October 2000; pp. 1633–1641.
13. Lim, Y.F. Cellular bucket brigades. *Oper. Res.* **2011**, *59*, 1539–1545. [CrossRef]
14. Lim, Y.F. Performance of cellular bucket brigades with hand-off times. *Prod. Oper. Manag.* **2017**, *26*, 1915–1923. [CrossRef]
15. Anderson, C.; Boomsma, J.J.; Bartholdi, J.J. Task partitioning in insect societies: Bucket brigades. *Insectes Sociaux* **2002**, *49*, 1–10. [CrossRef]
16. Bartholdi, J.J.; Eisenstein, D.D. A production line that balances itself. *Oper. Res.* **1996**, *44*, 21–34. [CrossRef]
17. Koo, P.H. The use of bucket brigades in zone order picking systems. *OR Spectr.* **2009**, *31*, 759–774. [CrossRef]
18. Sahin, M.; Kellegoz, T. A new mixed-integer linear programming formulation and particle swarm optimization based hybrid heuristic for the problem of resource investment and balancing of the assembly line with multi-manned workstations. *Comput. Ind. Eng.* **2019**, *133*, 107–120. [CrossRef]
19. Dimitriadis, S.G. Assembly line balancing and group working: A heuristic procedure for workers' groups operating on the same product and workstation. *Comput. Oper. Res.* **2006**, *33*, 2757–2774. [CrossRef]
20. Süer, G.A.; Sinaki, R.Y.; Sadeghi, A. A Hierarchical hybrid heuristic-optimization approach for multi-product assembly line design problem. *Procedia Manuf.* **2019**, *39*, 1067–1075. [CrossRef]
21. Rabbani, M.; Mahmood, S.; Manavizadeh, N. Mixed model U-line balancing type-1 problem: A new approach. *J. Manuf. Syst.* **2012**, *31*, 131–138. [CrossRef]
22. Li, Z.; Kucukkoc, I.; Zhang, Z. Branch, bound and remember algorithm for U-shaped assembly line balancing problem. *Comput. Ind. Eng.* **2018**, *124*, 24–35. [CrossRef]
23. Chiang, W.-C.; Urban, T.L. The stochastic U-line balancing problem: A heuristic procedure. *Eur. J. Oper. Res.* **2006**, *175*, 1767–1781. [CrossRef]
24. Lopes, T.C.; Michelsa, A.S.; Sikora, C.G.S.; Molina, R.G.; Magatãoa, L. Balancing and cyclically sequencing synchronous, asynchronous, and hybrid unpaced assembly lines. *Int. J. Prod. Econ.* **2018**, *203*, 216–224. [CrossRef]
25. Li, Z.; Janardhanan, M.N.; Tang, Q.; Ponnambalam, S.G. Model and metaheuristics for robotic two-sided assembly line balancing problems with setup times. *Swarm Evol. Comput.* **2019**, *50*, 100567. [CrossRef]

26. Bartholdi, J.J.; Eisenstein, D.D.; Foley, R.D. Performance of bucket brigades when work is stochastic. *Oper. Res.* **2001**, *49*, 710–719. [CrossRef]
27. Bartholdi, J.J.; Eisenstein, D.D. Using bucket brigades to migrate from craft manufacturing to assembly lines. *Manuf. Serv. Oper. Manag.* **2005**, *7*, 121–129. [CrossRef]
28. Munoz, L.F.; Villalobos, R. Work allocation strategies for serial assembly lines. *Int. J. Prod. Res.* **2002**, *40*, 1835–1852. [CrossRef]
29. Armbruster, D.; Ge, E.S. Bucket brigades revisited: Are they always effective? *Eur. J. Oper. Res.* **2006**, *172*, 213–229. [CrossRef]
30. Hirotani, D.; Myreshka, K.; Morikawa, K.; Takahashi, K. Analysis and design of self-balancing production line. *Comput. Ind. Eng.* **2006**, *50*, 488–502. [CrossRef]
31. Carlo, F.D.; Borgia, O.; Tucci, M. Bucket brigades to increase productivity in a luxury assembly line. *Int. J. Eng. Bus. Manag.* **2013**, *5*, 28. [CrossRef]
32. Hong, S.; Kim, Y. The effects of loosely coupled hand-off operations on bucket brigade order picking systems. *Ind. Eng. Manag. Syst.* **2018**, *17*, 745–756. [CrossRef]
33. Hong, S.; Johnson, A.L.; Peters, B.A. Quantifying picker blocking in a bucket brigade order picking system. *Int. J. Prod. Econ.* **2015**, *170 Pt C*, 862–873. [CrossRef]
34. Pratama, A.T.; Takahashi, K.; Morikawa, K.; Nagasawa, K.; Hirotani, D. Cellular bucket brigades with worker collaboration on U-lines with discrete workstations. *Ind. Eng. Manag. Syst.* **2018**, *17*, 531–549. [CrossRef]
35. Segal, M.; Whitt, W. A queueing network analyzer for manufacturing. *Teletraffic Sci. New Cost-Eff. Syst. Netw. Serv.* **1989**, *ITC-12*, 1146–1152.
36. Hong, S. The effects of picker-oriented operational factors on hand-off delay in a bucket brigade order picking system. *OR Spectr.* **2018**, *40*, 781–808. [CrossRef]
37. Hopp, W.J.; Spearman, M.L. *Factory Physics*, 2nd ed.; Irwin/McGraw-Hill: Boston, MA, USA, 2000.
38. Schultz, K.L.; McClain, J.O.; Thomas, L.J. Overcoming the dark side of work flexibility. *J. Oper. Manag.* **2003**, *21*, 81–92. [CrossRef]
39. Cesani, V.I.; Steudel, H.J. A study of labor assignment flexibility in cellular manufacturing systems. *Comput. Ind. Eng.* **2005**, *48*, 571–5915. [CrossRef]

Article

Design Model of Flow Lines to Include Switch-Off Policies Reducing Energy Consumption

Paolo Renna * and Sergio Materi

School of Engineering, University of Basilicata, 85100 Potenza, Italy; sergio.materi@unibas.it
* Correspondence: paolo.renna@unibas.it; Tel.: +39-0971-205143

Received: 27 January 2020; Accepted: 18 February 2020; Published: 21 February 2020

Abstract: One of the most promising approaches to reduce the amount of energy consumed in manufacturing systems is the switch off policy. This policy reduces the energy consumed when the machines are in the idle state. The main weakness of this policy is the reduction in the production rate of the manufacturing systems. The works proposed in the literature do not consider the design of the production lines for the introduction of switch off policies. This work proposes a design model for production lines that include a targeted imbalance among the workstations to cause designed idle time. The switch-off policy introduced in such designed production lines allows for a reduction in the energy consumed with any production rate loss. Simulation tests are conducted to verify the benefits of switch off policies in production lines designed for its. The simulation results show that the proposed line design allows for a reduction in energy consumption, with a defined loss in the throughput. The application of switch-off policies in the proposed flow line leads to a significant reduction in the energy used in unproductive states controlling the production loss.

Keywords: flow-line design; switch-off policy; energy consumption

1. Introduction and Motivations

Nowadays, the industrial manufacturers study energy-efficient models because of the costs and environmental impact of energy consumption [1,2].

The manufacturing activities are characterized by a relevant demand from around 40% with relevant growth expectations between 2018 and 2050 across all cases. Increases in industrial energy use from increasing shipments are partially offset by efficiency gains [3].

Studies by the European Association of the Machine Tool Industries [4,5] on the discrete part manufacturing discussed the importance and environmental impacts of electrical energy consumption.

The main models proposed in the literature, for the design of manufacturing systems, focused primarily on the performance in regards to productivity, quality, and work in process, etc. Recently, the models proposed include the energy efficiency issue, but Gahm et al. [6] emphasized how the scheduling models that include energy-saving may reduce the other goals of the manufacturing systems. Then, it is more important to propose a model that reduces energy consumption without reducing the productivity performance of manufacturing companies.

The works proposed regarding machining energy consumption [7,8] identified three main energy shares: start-up operations (computers and fans, coolant pumps, etc.), runtime operations (tool change, Jog axis, etc.) and material removal operations (machining). The first and second parts are constant and independent of the operation, while the third part is variable and depends on the machining operation.

Gutoski et al. [8–10] underlined how the constant independent of the operation is the major part of the energy consumed by the machining machines and this trend has grown in past years.

The reduction in the energy consumption of these parts can be obtained without studying the effect on the particular machining operation as: roughness, cutting paths, cutting time, etc. Therefore, it is more simply to extend a model that reduces the constant energy to several machining systems and other production systems.

A promising approach proposed in the literature to reduce the constant energy consumed by the machining machines is the switch-off policy [11]. Reducing the idle time is a good approach to realize energy-efficient production [12]. One example is the suppliers for small-parts in aircrafts, where the idle periods accounted for 16% of the total production time, and about 13% of the total energy consumption could be saved if the idle machines were switched off [13]. Another study by Weinert and Mose [14] highlighted how almost 50% to 60% of energy consumption could be reduced by turning the standby machines into energy-saving states.

The switch-off policies switch off when the machine is in the idle state, which reduces the constant energy consumed during this state. When there are a determined number of parts waiting for the operation, the machine switches on, subject to a warm-up period to be operative. The decision of the switch on is crucial to maintain the productivity level of the production line. Some approaches have been proposed, by Frigerio and Matta [15,16], with a significant reduction in energy consumption in a flow lines context. The switch-off models proposed works on a flow line designed following classical objectives that solve the simple assembly line balancing problem [17,18]. An extensive survey of the flow line production system balancing was discussed in References [19,20].

Mathematical models to support the decision towards the switch off/on, which works off-line, have been proposed by Mashaei and Lennartson [21] and Jia et al. [22]. These models did not handle the uncertainty of the production system and the optimization can lead to high computational complexity when the number of machines is higher; this reduces the potential application in real industrial manufacturing systems.

Su et al. [23] and Frigerio and Matta [16] studied upstream, downstream, and mixed policies in a single machine, and the extension to a production line, assuming stochastic arrivals, and constant warm-up under to stock control. These models did not consider how the control of the machines may reduce the production performance of the manufacturing system.

Renna [24] studied a dynamic and adaptive control strategy to switch-off the machines on a production line under the pull control policy. The policy proposed uses the information of the buffers and the level of customer satisfaction. The results also underlined the potential application of the switch-off policies in production lines under pull control.

Duque et al. [25] studied one machine with one buffer with a fuzzy controller that includes the information about the buffer, the machine state and the production rate required, considering the warm-up energy. The controller was tested through simulation experiments and it was observed that a large amount of energy could be saved without affecting the throughput significantly.

Wang et al. [26,27] proposed a method based on a fuzzy controller that includes the information of the upstream, downstream buffer and the status of the machine. They proposed a set of fuzzy rules to take the decision to the switch off/on policy. The simulation experiments show that the proposed approach can be a simple, practical way towards the energy saving operation with accepted throughput loss.

Marzano et al. [28] proposed a model that controls the machine on-line acquiring data and estimating the risk of the control policy actions. The model has several limitations because it is tested for a specific distribution for the estimation of the parameters and the tests are conducted considering a production line composed of two machines.

The discussion of the past works highlights the following issues:

- The switch-off policies are mainly proposed for a single machine and some works considered a flow line with the introduction of buffers;
- Few works evaluate the reduction of the performance level due to the introduction of the switch-off policy, but only energy consumption;
- Works were proposed to consider the possibility of a switch-off policy and were considered from the design step of the production line.

In response to the limits in the literature, this paper proposed a design model that includes the switch-off policy by first asking:

RQ1: what is the impact of the design model proposed on the performance of the production line in terms of energy saving maximizing the production rate?

Some previous works show how the switch-off policy causes production loss, then our second research question asks:

RQ2: can the constraint of a limited reduction loss improve significantly the energy saving of the production line obtaining an adequate trade off?

The research proposed concerns about the development of a design model of the flow lines that includes the possibility of the switch-off already in this design step. Then, the switch-off policy is introduced in the flow line designed by the proposed model. The proposed design and switch-off policy is compared to a flow line designed with the classical objectives with the same switch-off policy.

Simulation experiments will be used to answers our two research questions. The simulations were conducted by considering different levels of production rates in the design model and different levels of the buffers (upstream and downstream) that control the switch-off policy.

The paper is organized as follows. Section 2 described the design model of the flow line, while Section 3 introduces the switch-off policy used. Section 4 introduces the reference context investigated with simulation scenarios, while Section 5 discussed the numerical results. Section 6 provides the conclusions and a future research path.

2. Flow Line Design Model

The problem deals with the designing of a flow line composed by M stations that manufactures one product type. The product consists of N operations to process; these operations should be assigned to the M workstations, following the precedence constrains. The variables of the model are the assignments of the operations to the workstations of the flow line. The operation assignment binary variable x_{ij} is defined as follows (Equation (1)):

$$x_{ij} = \begin{cases} 1 & if\ the\ i-th\ operation\ is\ assigned\ to\ the\ j-th\ station \\ 0 & otherwise \end{cases} \tag{1}$$

The precedence constrains binary variable ensures that the *i-th* operation must be completed before the *k-th* operation, and it is computed as (Equation (2)):

$$v_{ik} = \begin{cases} 1 & if\ the\ i-th\ operation\ preceds\ the\ k-th\ station \\ 0 & otherwise \end{cases} \tag{2}$$

The processing time T_j of the *j-th* station is the sum of the processing time of the operations assigned to *j-th* station, as follows (Equation (3)):

$$T_j = \sum\nolimits_i^N t_i x_{ij} \tag{3}$$

The cycle time C of the production line is equal to the maximum of the stations' processing times, as shown in Equation (4)

$$C = MAX_j\{T_j\} \tag{4}$$

The *j-th* station idle time is defined as follows (Equation (5)):

$$T_{d,j} = C - T_j \tag{5}$$

The idle time for each cycle is given by Equation (6):

$$TT_d = \sum_{j=1}^{M} \left(T_{d,j}\right) \tag{6}$$

The difference of processing time between nearby stations is called distance d_j and it is defined as follows (Equation (7)):

$$d_j = T_{j+1} - T_j \tag{7}$$

A positive value of the distance d_j forms a couple of workstations when the first has a higher velocity than the second workstation. Then, the first workstation can fill the downstream buffer and goes into the off state, reducing energy consumption. Therefore, the flow line consists of a couple of workstations to facilitate the off state of the first workstation of the couple. Figure 1 shows the concept of the distance of a couple of workstations. In the following figure, the term WS_j has been used to identify the j-th workstation; the distance between two stations has been obtained as the difference of the processing time of the second workstation (with higher working time) and the first workstation of the couple. In Figure 1, the processing time is linked to the stations with, respectively, lower and higher productivity per couple, as represented in blue and green.

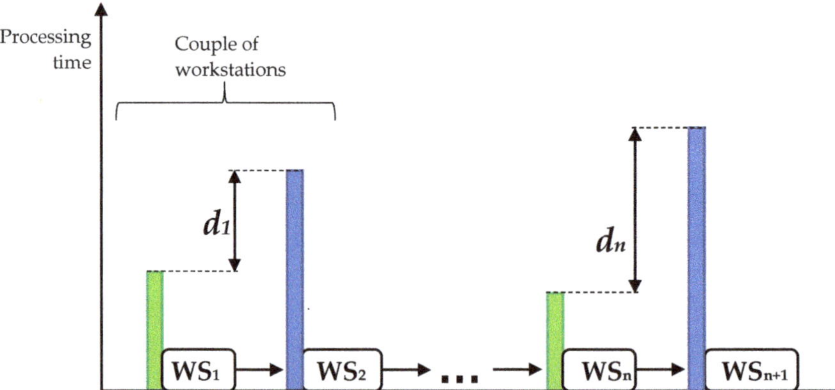

Figure 1. Distance between stations.

To achieve an unbalanced flow line, the objective function is achieved by maximizing the sum of the distance between stations (Equation (8)):

$$MAX\left(\sum\nolimits_j d_j\right) \quad with\ j = 1, 3, 5, \ldots \tag{8}$$

This is subject to the following constrains (Equations (9)–(11))

$$\sum\nolimits_{j=1}^{M} x_{ij} = 1 \qquad with\ i = 1, \ldots, N \tag{9}$$

$$\sum\nolimits_{j=1}^{M} j \ \ x_{ij} \leq \sum\nolimits_{j=1}^{M} j\ x_{kj} \qquad \forall\ v_{ik} = 1 \tag{10}$$

$$\sum\nolimits_{i=1}^{N} t_i \ \ x_{ij} \leq C_{max} \qquad with\ j = 1, \ldots, M \tag{11}$$

Equation (9) ensures that each operation is only assigned to one machine. Equation (10) ensures that the constraints on the precedence of operations are respected. Finally, Equation (11) ensures that the station processing times are below the maximum fixed cycle time. By setting the cycle time, maximizing the objective function achieves an unbalanced line that respects the targeted productivity. Equation (8) results in an unbalanced flow line with high idle times between sequential stations. In order to achieve more downtime, and therefore higher energy saving, this research proposes a different objective function from past literature, where the focus instead is on obtaining the minimum cycle time [17], the minimum number of stations, and the minimum idle time [18].

Figure 2 shows the framework used for the design of unbalanced flow lines. First, using the mathematical model, the flow line with maximum productivity and minimum cycle time has been achieved. The obtained maximum productivity flow line and the minimum cycle time has been calculated as the bottleneck station processing time. The minimum cycle time has been used as a parameter to design unbalanced flow lines using the maximization of the sum of distances as the objective function according Equation (8). Then, increasing and fixing the cycle time, the maximization of the sum of the distance results in several unbalanced flow lines (in this paper, three unbalanced flow lines have been obtained). Then, the simulation will be studied if the increment of the cycle time can lead to important energy reduction.

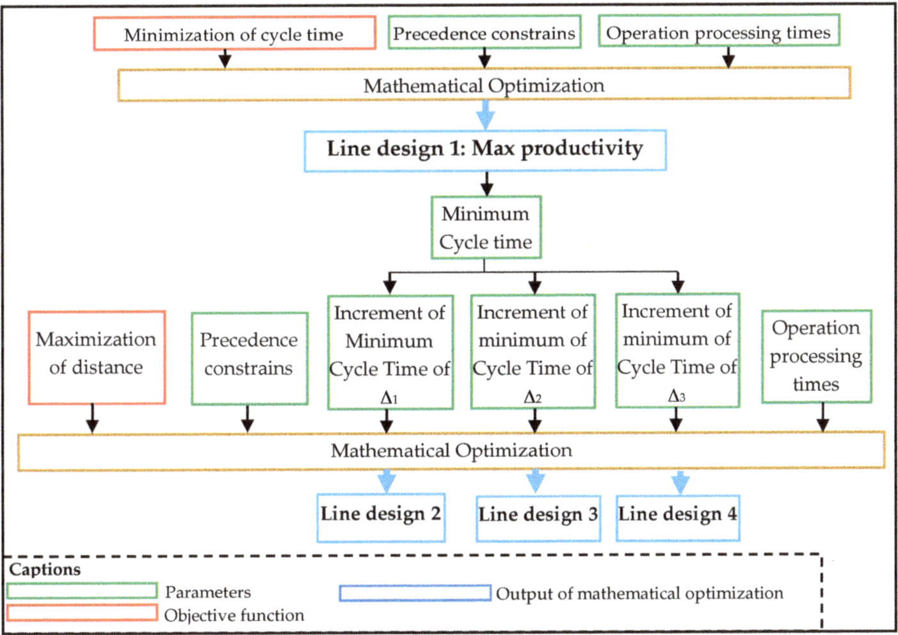

Figure 2. Framework for flow line designs.

3. Switch off Policy

As described in the literature [23], switching off policies based on buffer level information can lead to significant energy savings without reducing productivity. The proposed policies are upstream (UP), downstream (DP), and upstream and downstream (UDP). In the upstream policy, the machine switches off when the upstream buffer is empty and switches on when the upstream buffer level is $N^{U}{}_{on}$. The level of downstream buffer controls the state in the downstream policy. The machine switches off when the threshold $N^{D}{}_{off}$ is reached and turns on when the number of pieces in the buffer is equal to $N^{D}{}_{on}$. According to Reference [15], the state of machine s_j is defined as follows (Equation (12)):

$$s_j = \begin{cases} 1 & if\ out-of-service \\ 2 & if\ idle \\ 3 & if\ in\ start-up \\ 4 & if\ working \end{cases} \tag{12}$$

The states 1, 2 and 3 are unproductive, i.e., no pieces are being processed when a machine is in one of these states. The states 1 and 2 are called inactive states. According to Reference [23], the upstream and downstream combines the UP and DP policies as follows (Equation (13)):

$$\begin{cases} Switch-Off & if\ s_j = 2\ AND\ (n_j = 0\ OR\ n_{j+1} \geq N^D_{off}) \\ Switch-On & if\ s_j = 1\ AND\ (n_j = N^U_{on}\ AND\ n_{j+1} \leq N^D_{on}) \end{cases} \tag{13}$$

Figure 3 summarizes the states of the generic machine and the transition from one state to another.

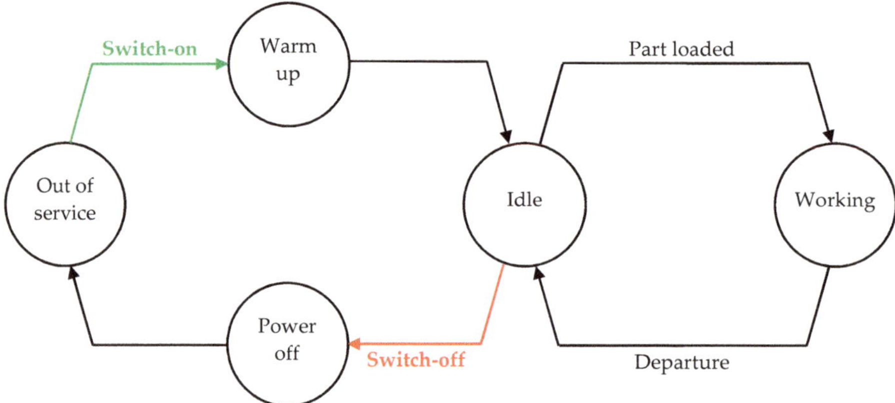

Figure 3. Machine states.

4. Reference Context and Simulation Scenarios

Using the mathematical model described in Section 2, four production lines with 10 stations and 20 tasks to complete have been designed. The flow line only produces one product type.

The operation processing times and the precedence constraints are described in Table 1 and have been obtained from the simple assembly line balancing problem dataset (SALBP) according to Reference [29].

Table 1. Operation processing time [s].

Operations	Processing Time (s)	Precedence	Operation	Processing Time (s)	Precedence
1	142	-	11	97	7
2	34	-	12	132	8
3	140	-	13	107	10, 11
4	214	-	14	132	12
5	121	-	15	69	12
6	279	1	16	169	13
7	50	2	17	73	13
8	282	4	18	231	13
9	129	5	19	120	15
10	175	6	20	186	14

The basic assumptions for the line design are the following:

- The stations can perform every possible operation assigned by the design model;
- The precedence constraints are fixed;
- No machine failure has been considered;
- The operations processing times are deterministic and are the same for each station.

Figure 4 shows the precedence graph. In the precedence graph, each task is represented as a node, and each direct precedence constraint is illustrated as an arrow that links node i and k if the i-th operation must precede the k-th operation. For example, in Figure 4, operation 6 must be executed before operation 10.

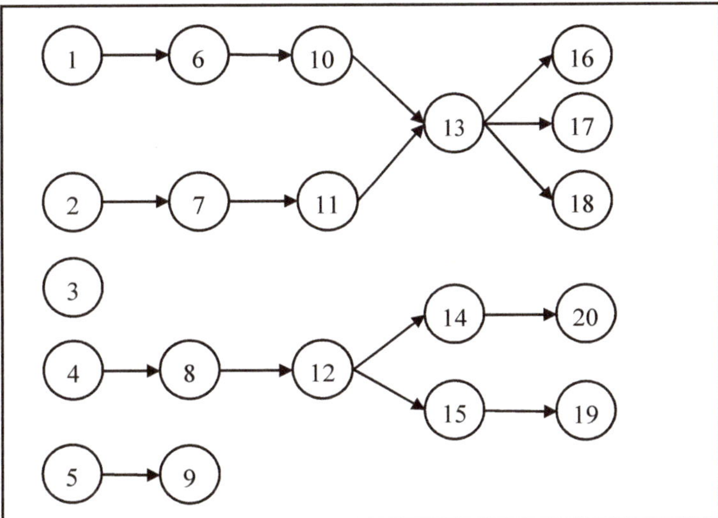

Figure 4. Precedence graph.

Four flow lines are obtained in order to get respectively:

1. Minimization of total idle time for each cycle with minimum Cycle Time ($MinTT_d$);
2. Maximization of distance between stations with a 2.5% increment of minimum Cycle Time (2.5% loss production rate, MaxD_2.5%);
3. Maximization of distance between stations with a 5% increment of minimum Cycle Time (5% loss production rate, MaxD_5%);
4. Maximization of distance between stations with a 10% increment of minimum Cycle Time (10% loss production rate, MaxD_10%).

The solution of the mathematical model, considering the previous constrains, gives the following flow line designs (Table 2); moreover, the station processing times and the station idle times have been reported in Table 2.

The first production line is considered to be the benchmark of the other production lines, because it gives a balanced production line with 10 stations and the minimum cycle time.

A discrete event simulation, implemented in Arena, has been used to evaluate the performances of the four flow line designs and to analyze the application of switch off policies in unbalanced production lines.

Table 2. Stations processing and idle times [s].

	MinTT$_d$		MaxD_2.5%		MaxD_5%		MaxD_10%	
Station	**Station Processing Time [s]**	**Station Idle Time [s]**	**Station Processing Time [s]**	**Station Idle Time [s]**	**Station Processing Time [s]**	**Station Idle Time [s]**	**Station Processing Time [s]**	**Station Idle Time [s]**
1	298	6	248	63	248	70	214	119
2	282	22	311	0	313	5	282	51
3	272	32	282	29	282	36	263	70
4	274	30	298	13	298	20	333	0
5	279	25	274	37	279	39	279	54
6	287	17	306	5	315	3	329	4
7	304	0	279	32	236	82	269	64
8	293	11	304	7	304	14	333	0
9	304	0	276	35	289	29	276	57
10	289	15	304	7	318	0	304	29
TT_d		158		228		298		448

Each model has been simulated by considering machines in the "always on" (AO) state and the UPD switch off policy.

The basic assumptions of the AO model are:

- Each station has a buffer;
- The buffer capacity is fixed, and equal to K;
- The buffer of the first station is always full, that is the raw material is always available;
- The power absorbed in each state is equal for all machines.

In addition, for the models with switch off policies, the following conditions apply:

- Each station is controlled by a switch off policy;
- The control policy parameters ($N^U{}_{on}$, $N^D{}_{off}$, $N^D{}_{on}$) are the same for stations from 2 to 9;
- The first station has only DP policy;
- The last station has only UP policy.

As described by Frigerio and Matta [15], the production line machines can be in the following states:

- Working state: the machine is processing a piece and absorbs the power P_w;
- Idle state: the machine is ready to work a part, and absorbs the power P_i;
- Out-of-service (Inactive) state: the machine is not ready to process a part. In this state the machine absorbs the minimum amount of power P_{off};
- Warmup state: the machine changes its state from Out-of-service in idle or working state, consuming the power P_{wu} for the time to complete the warmup t_{wu}.

According to [23], the power required by a generic machine in each state is:

- $P_w = 12$ kW;
- $P_i = 5.35$ kW;
- $P_{off} = 0.52$ kW;
- $P_{wu} = 6$ kW for $t_{wu} = 20$ s.

To determine the best switch off control parameters, a full factorial design has been developed. The factors considered are $N^U{}_{on}$, $N^D{}_{off}$, $N^D{}_{on}$, and three levels for each factor are evaluated as follows:

- $N^U{}_{on} = [1, 2, 3]$;
- $N^D{}_{on} = [4, 5, 6]$;
- $N^D{}_{off} = [7, 8, 9]$;

According to Reference [23], the levels of buffer to switch off or switch on machines respect the following constraints (Equations (14)–(16)):

$$N^D_{off,j} \geq N^D_{on,j} \quad (14)$$

$$N^D_{off,j} \geq N^U_{on,j+1} \quad (15)$$

$$N^D_{on,j} \geq N^U_{on,j+1} \quad (16)$$

Figure 5 reports on the experiment results for model 1. The design with the lowest inactive time has a value of 0, the design with the highest inactive time has a value of 1. These results are used to set these parameters for the simulation tests.

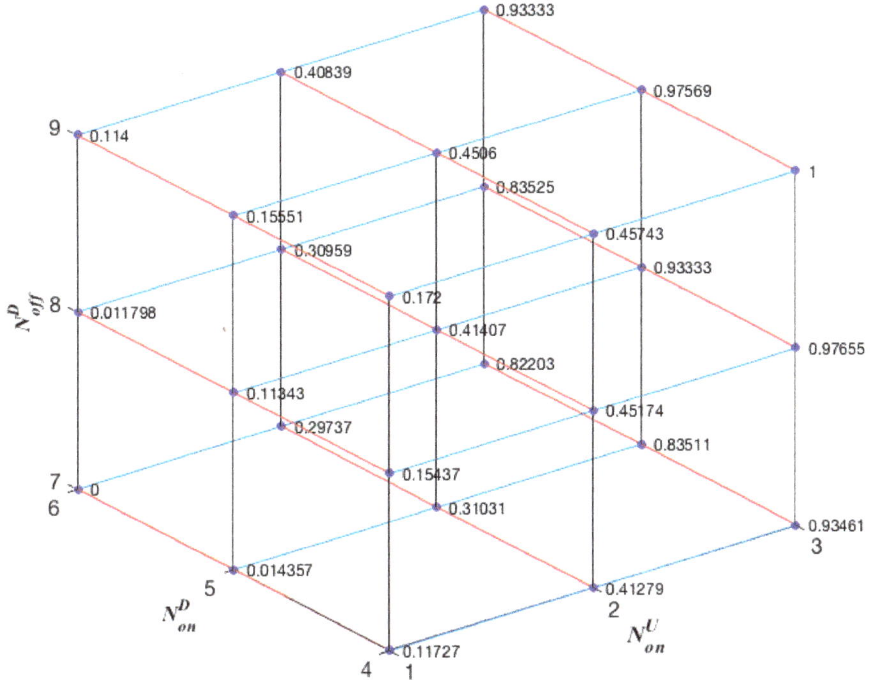

Figure 5. Experiments results for line design $MinTT_d$.

The results of the other experiments for models MaxD_2.5%, MaxD_5%, and MaxD_10% are reported in Appendix A.

As shown in Figure 2, and in the figures (Figures A1–A3) in Appendix A, for all the four design models, the set $N^U{}_{on} = 3$, $N^D{}_{on} = 4$, $N^D{}_{off} = 9$, gives the maximum time spent in the idle and inactive state. Therefore, the simulation scenarios and the evaluation of the performances have been obtained by considering the best control parameters. According to Reference [23], the discrete event simulation length is 10^7 s, and the initial transient is 10^5 s.

Figure 6 shows the setting of simulation scenarios and performance evaluations. Using mathematical optimization, the flow line design and the operations assignment to the stations have been achieved. Using discrete event simulation, the best set of switch off-parameters have been obtained. Finally, by employing this set, the performances of the four lines have been evaluated and compared.

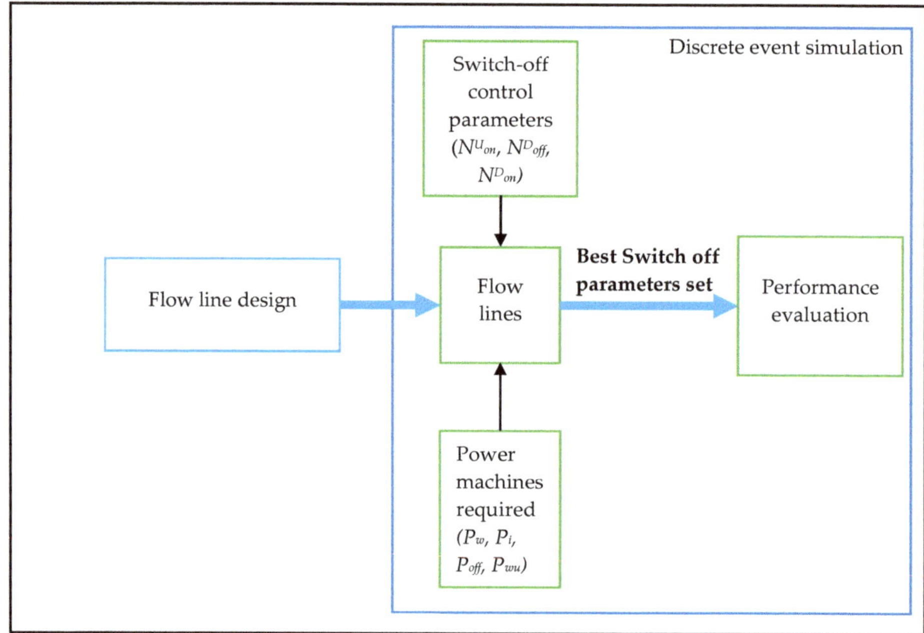

Figure 6. Simulation process and performances evaluation.

5. Numerical Results

As shown in the following figure (Figure 7), an increment of the cycle idle time for each station has been achieved by choosing the maximization of the distance between stations as the objective function, instead of the minimization of idle time. The maximum increment of TT_d has been obtained in the case where the cycle time has been increased by 10%.

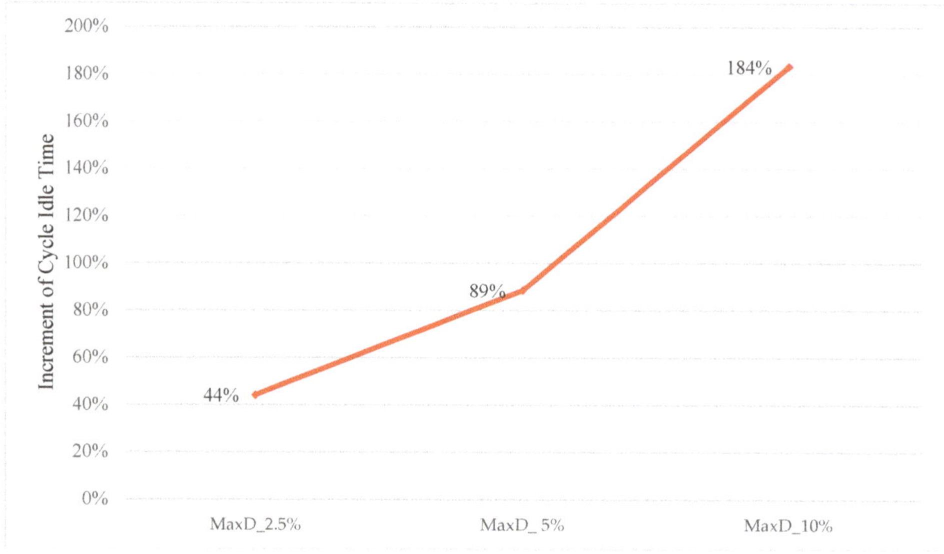

Figure 7. Increment of idle time for each cycle.

The discrete event simulation results, as compared to the model $MinTT_d$, are shown in Figure 8. The results in the figure only consider the effects of the unbalanced line on the idle time, so the simulations have been made by considering machines with an always on control policy. It can be noticed that the higher the idle time, the lower the throughput. Indeed, by increasing the distance, stations are in the idle state for longer than in a balanced flow line. Increased cycle time allows for mathematical optimization, which aims to maximize the distance between station process times and to have greater freedom in assigning operations to the machines. The 10% increase in cycle time (MaxD_10%) results in a flow line where the idle time has grown by 159%, but productivity has decreased by only 9%.

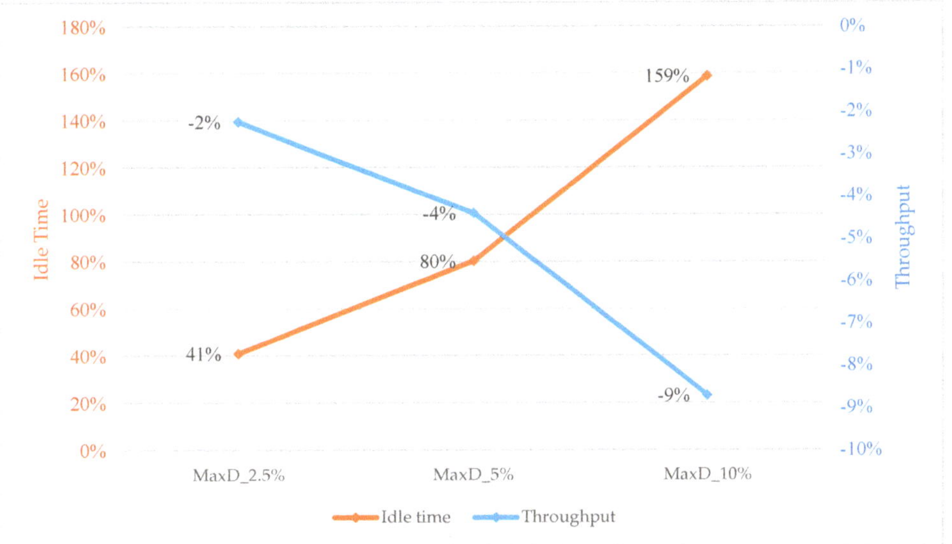

Figure 8. Results with always on policy.

However, the increase in downtime leads to more energy consumption in a non-productive state. In order to reduce energy consumption, switch-off policies should be in place. The UDP policy achieves a significant reduction in energy consumption in non-productive states. In Figure 9, the energy consumption in unproductive states (Idle, Out-of-service, Warmup) of the four lines ($MinTT_d$, MaxD_2.5%, MaxD_5%, MaxD_10%) are compared to a case with machines that are always on and those applying switch off policies. In all the cases analyzed, adopting a shutdown policy allows for a significant reduction in energy consumption in non-productive states, ranging from 86% to 89%.

For these reasons, switch-off policies in unbalanced flow lines are necessary and achieve a reduction in unproductive state energy consumption. If the machines are in always on states, the unbalancing flow line leads to a reduction in throughput and energy consumption due to more time being spent in the idle state. Staying for a long time in the idle state is detrimental, since in this state, the machine is ready to work and then absorbs power without producing. The switch-off policies warrant energy saving due to the lower energy consumption during warmup.

Designing an unbalanced flow line and controlling the machines state with a switch-off policy can lead to a reduction in total energy consumption, not only in unproductive states. Figure 10 shows the increment of energy consumption for an unbalanced flow line design with switch-off policies, compared to an always on balanced flow line. For this reason, designing a flow line to achieve a high unbalance under cycle time constraints, and applying a machine shutdown policy, leads to a reduction in total energy consumption.

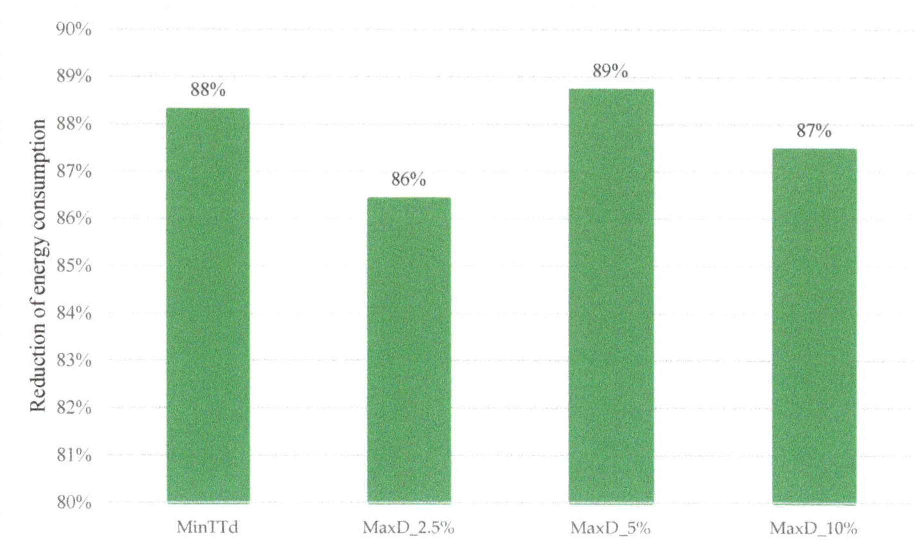

Figure 9. Reduction of energy consumption in unproductive states using switch-off policies.

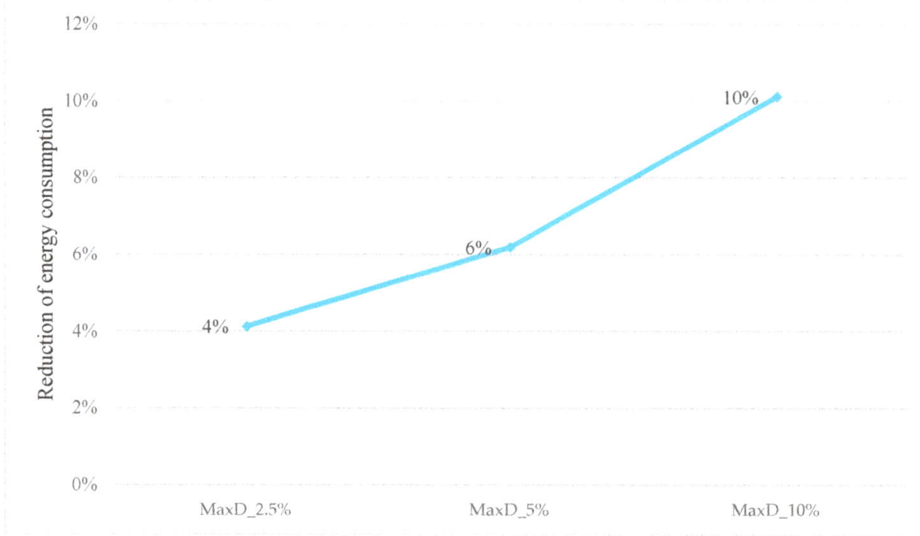

Figure 10. Total energy reduction.

Figure 11 shows the total time that stations hold in the inactive state and the number of warmups.

Like the number of pieces in the buffer, the number of times that machines switch on depends on the position of the bottleneck. Machines turn on fewer times when using model MaxD_5% than model Max_2.5%, even if the total inactive time is longer, because the last station is the bottleneck. It can be seen that, if the bottleneck is among the first machines on the line, it can lead to a reduction of the work in process and storage costs. However, this leads to more switch-ons and higher energy consumption during warmup.

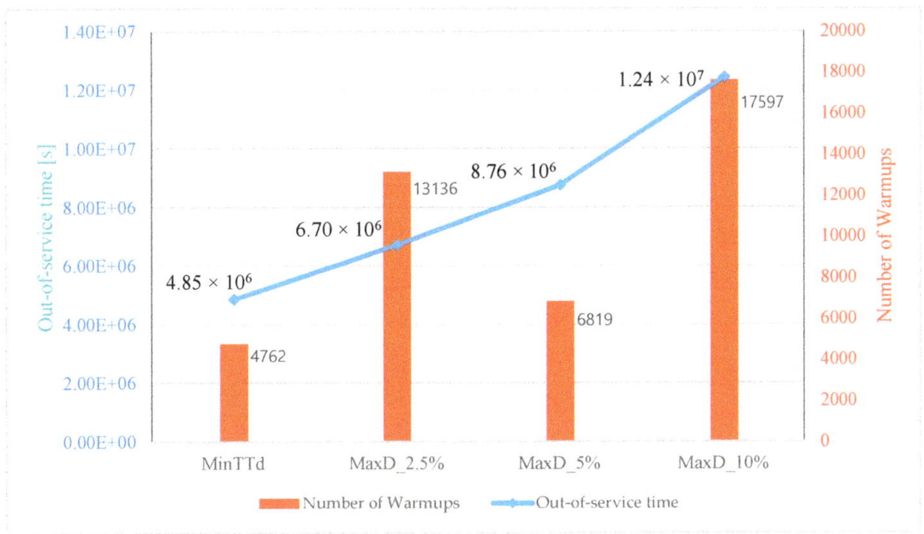

Figure 11. Warmup and out-of-service time.

Figure 12 reports the switch-off mean times. It can be noticed that the model MaxD_5% gives the maximum value of the mean inactive time. Thus, this configuration leads to a high inactive time for each switch off. Instead, in the model MaxD_2.5%, the mean time in the inactive state is the lowest; the machines turn into off-state since there are only few pieces in the buffer. The UP policy achieves a reduction in the machine energy consumption removing resource starvation. Indeed, the machine turns off when the upstream buffer is empty and then does not wait in the idle state using a high quantity of energy.

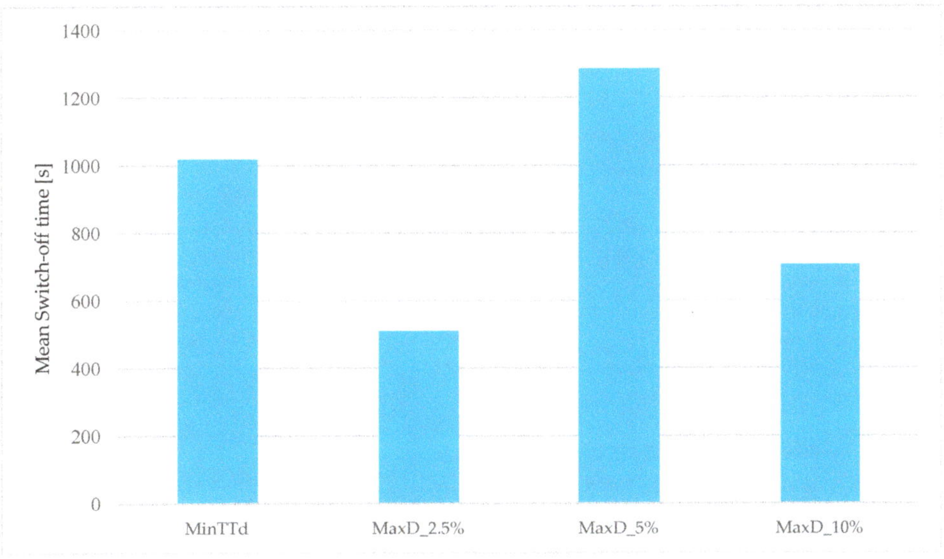

Figure 12. Mean switch off time for the four lines design with UPD policies.

The following figure (Figure 13) shows the mean of the pieces in the buffers of the four models with the switch off policy. It can be noticed that the number of pieces in the buffers does not depend on the chosen objective function. Indeed, the number of elements in the buffers depends on the position of the bottleneck. In fact, in model MaxD_2.5%, as written in Table 2, the bottleneck is the second machine on the line. For these reasons, if the elements in the buffers are lower than $N^{U}{}_{on}$, then the UPD policy degrades for the machines below the bottleneck in the upstream policy. Future work may investigate the effect of changing the bottleneck position in an unbalanced flow line in order to obtain the best compromise between WIP (Work in Process) and the machines switch-off mean times.

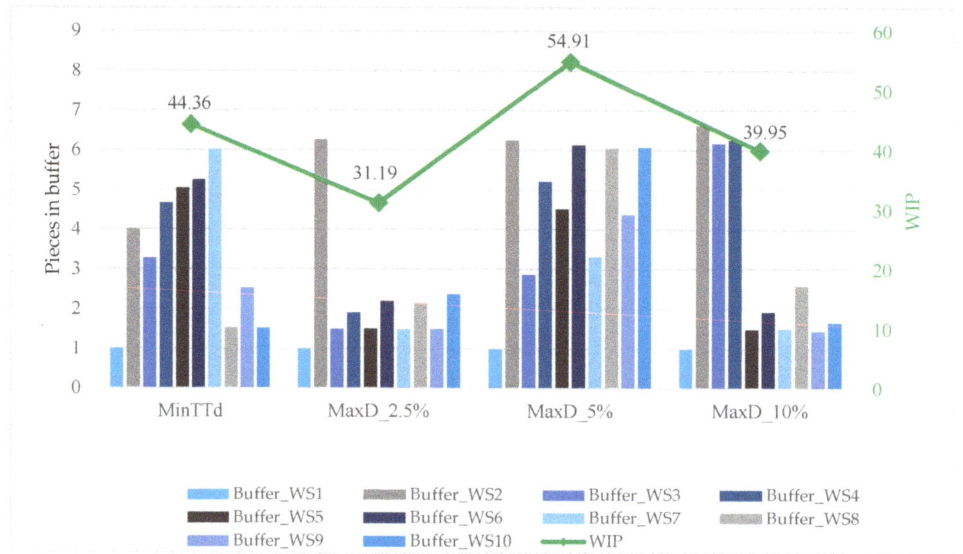

Figure 13. Pieces in buffer and mean of the work in progress.

6. Conclusions

A design model of flow lines to improve the efficiency of the switch-off policy has been proposed.

The model is based on the introduction of a couple of workstations; the first is more rapid than the second in order to facilitate the switch-off of the first machine. Moreover, the proposed model allows for the evaluation of the potential benefit of energy consumption with a determined reduction in the production rate. In unbalanced flow lines, the higher idle times lead to a reduction in the total energy required for the production. However, in the idle state, stations absorb energy without working. For these reasons, applying the switch-off policy in an unbalance flow line is necessary to reduce energy consumption in the unproductive state.

The design model is used with the discrete event simulation to highlight the energy saving with the different levels of production loss fixed. In response, our first research question asked: what is the impact of the design model proposed on the performance of the production line in terms of energy saving maximizing the production rate?

Using the simulation, we have demonstrated that the model proposed could improve the reduction of energy consumption of the flow line more than the design model that does not consider the possibility of introducing the switch-off policy.

However, in response to our second research question: can the constraint of a limited reduction loss improve significantly the energy saving of the production line obtaining an adequate trade off?

Our results have demonstrated that the model can support the decision about the better trade-off between the production rate level and energy consumption reduction. Moreover, the results highlighted the better choice, if the objective is also the reduction in the number of on/off activities that can affect the maintenance of the machines. The results show that the number of warmups can be reduced by properly choosing the bottleneck position, respecting the precedence constraints, in order to achieve further energy saving.

From a managerial viewpoint, the study was motivated by the important issue of energy consumption of flow lines. Our results suggests that: (i) the design model should be adapted to introduce the switch-off policy to obtain a higher benefit from the switch-off policies; (ii) it is possible to evaluate, with the use of the simulation, the effect of a targeted reduction of production rate (for example in a determined production planning period) to improve the energy consumption reduction; and (iii) the model proposed can be extended to different flow lines to support the decision making about the design and potential energy reduction.

A limitation of our study is that the machines were not affected by failures and the processing times are deterministic. Machine failures have a significant impact on productivity. For this reason, integrating maintenance during scheduled machine downtime can lead to a reduction in production losses. Therefore, future works will consider the unbalance flow line design in a pull system that also considers machine failures and stochastic processing times. Furthermore, different switch-off policies, for example, that also consider demand fluctuations for shutdowns, and their effect on throughput and on energy consumption, will be investigated.

Author Contributions: S.M. performed the experiments and wrote the paper, and P.R. designed the experiments and revised the paper. All authors have read and agreed to the published version of the manuscript.

Funding: This research received no external funding

Conflicts of Interest: The authors declare no conflict of interest.

Nomenclature

C	Cycle time
C_{max}	Maximum fixed cycle time
d_j	Distance between nearby stations
$i = 1,.., N$	Operation index
$j = 1,.., M$	Station index
K_j	j-th buffer capacity
n_j	Number of parts in j-th buffer
$N^D{}_{off}$	Downstream buffer level to switch off
$N^D{}_{on}$	Downstream buffer level to switch on
$N^U{}_{on}$	Upstream buffer level to switch on
P_i	Power in idle state
P_{off}	Power in out-of-service state
P_w	Power in working state
P_{wu}	Power in warm up state
s_j	State of the j-th station
$T_{d,j}$	j-th station idle time
t_i	i-th operation processing time
T_j	j-th station processing time
t_{wu}	Time in warm-up state
TT_d	Cycle idle time
v_{ik}	Precedence constrains binary variable
WS_j	j-th station
x_{ij}	Operation assignment binary variable

Appendix A

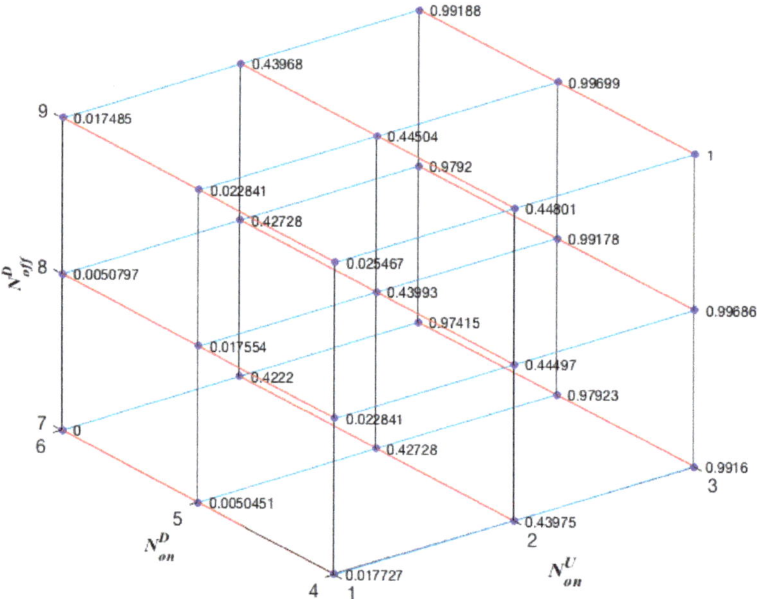

Figure A1. Experiments results for line design MaxD_2.5%.

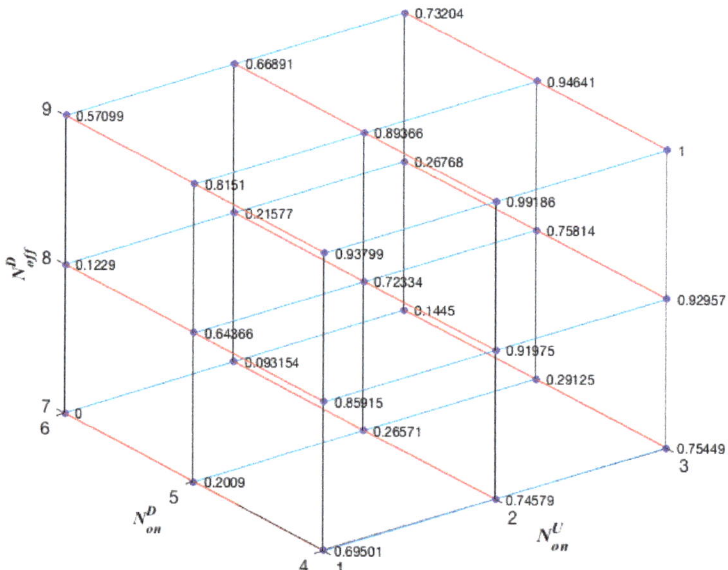

Figure A2. Experiments results for line design MaxD_5%.

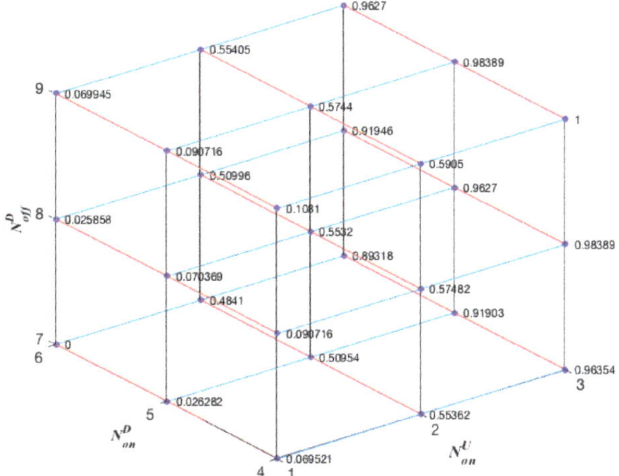

Figure A3. Experiments results for line design MaxD_10%.

References

1. The Cadmus Group. *Regional Electricity Emission Factors Final Report*; The Cadmus Group: Boston, MA, USA, 1998.
2. Deif, A.M. A system model for green manufacturing. *J. Adv. Prod. Eng. Manag.* **2011**, *19*, 1553–1559. [CrossRef]
3. U.S. Energy Information Administration. Annual Energy Outlook 2019. Available online: https://www.eia.gov/outlooks/aeo/pdf/aeo2019.pdf (accessed on 5 December 2019).
4. Cecimo. *Concept Description for CECIMO's Self-Regulatory Initiative (SRI) for the Sector Specific Implementation of the Directive 2005/32/EC*; Cecimo: Brussels, Belgium, 2009; Available online: https://www.eup-network.de/fileadmin/user_upload/Produktgruppen/Lots/Working_Documents/Lot_ENTR_05_machine_tools/draft_self_regulation_machine_tools_2009-10.pdf (accessed on 12 December 2019).
5. Cecimo. *The European Machine Tool Sector and the Circular Economy*; Cecimo: Brussels, Belgium, 2019; Available online: https://www.cecimo.eu/wp-content/uploads/2019/05/Circular-Economy-Report.pdf (accessed on 5 December 2019).
6. Gahm, C.; Denz, F.; Dirr, M.; Tuma, A. Energy-Efficient Scheduling in Manufacturing Companies: A Review and Research Framework. *Eur. J. Oper. Res.* **2016**, *248*, 744–757. [CrossRef]
7. Dahmus, J.B.; Gutowski, T.G. An Environmental Analysis of Machining. In Proceedings of the ASME 2004 International Mechanical Engineering Congress and Exposition, Manufacturing Engineering Materials Hand, Anaheim, CA, USA, 13–19 November 2004; pp. 643–652. [CrossRef]
8. Gutowski, T.G.; Dahmus, J.B.; Thiriez, A. Electrical Energy Requirements for Manufacturing Processes. In Proceedings of the 13th CIRP International Conference of Life Cycle Engineering, Leuven, Belgium, 31 May–2 June 2006.
9. Gutowski, T.G.; Dahmus, J.B.; Thiriez, A.; Branham, M.; Jones, A. A Thermodynamic Characterization of Manufacturing Processes. In Proceedings of the 2007 IEEE International Symposium on Electronics and the Environment, Orlando, FL, USA, 7–10 May 2007; pp. 137–142. [CrossRef]
10. Gutowski, T.G.; Branham, M.S.; Dahmus, J.B.; Jones, A.J.; Thiriez, A.; Sekulic, D.P. Thermodynamic Analysis of Resources Used in Manufacturing Processes. *Environ. Sci. Technol.* **2009**, *43*, 1584–1590. [CrossRef] [PubMed]
11. Li, W.; Zein, A.; Kara, S.; Herrmann, C. An Investigation into Fixed Energy Consumption of Machine Tools. In Proceedings of the 18th CIRP International Conference on Life Cycle Engineering—Glocalized Solutions for Sustainability in Manufacturing, Braunschweig, Germany, 2–4 May 2011; pp. 268–273. [CrossRef]
12. Zhang, X.P.; Cheng, X.M. Energy Consumption, Carbon Emissions, and Economic Growth in China. *Ecol. Econ.* **2009**, *68*, 2706–2712. [CrossRef]

13. Twomey, J.; Yildirim, M.B.; Whitman, L.; Liao, H.; Ahmad, J. *Energy Profiles of Manufacturing Equipment for Reducing Energy Consumption in a Production Setting*; Working Paper; Wichita State Univ.: Wichita, KS, USA, 2018.
14. Weinert, N.; Mose, C. Investigation of Advanced Energy Saving Stand by Strategies for Production Systems. *Procedia CIRP* **2014**, *15*, 90–95. [CrossRef]
15. Frigerio, N.; Matta, A. Energy-Efficient Control Strategies for Machine Tools with Stochastic Arrivals. *IEEE Trans. Autom. Sci. Eng.* **2015**, *15*, 50–61. [CrossRef]
16. Frigerio, N.; Matta, A. Analysis on Energy Efficient Switching of Machine Tool with Stochastic Arrivals and Buffer Information. *IEEE Trans. Autom. Sci. Eng.* **2016**, *13*, 238–246. [CrossRef]
17. Baybars, I. A Survey of Exact Algorithms for the Simple Assembly Line Balancing Problem. *Manag. Sci.* **1986**, *32*, 909–932. [CrossRef]
18. Scholl, A. *Balancing and Sequencing Assembly Lines*, 2nd ed.; Physica-Verlag Heidelberg: Heidelberg, Germany, 1999.
19. Boysen, N.; Fliedner, M.; Scholl, A. A Classification of Assembly Line Balancing Problems. *Eur. J. Oper. Res.* **2007**, *183*, 674–693. [CrossRef]
20. Eghtesadifard, M.; Khalifeh, M.; Khorram, M. A Systematic Review of Research Themes and Hot Topics in Assembly Line Balancing through the Web of Science within 1990–2017. *Comput. Ind. Eng.* **2020**, *139*, 106–182. [CrossRef]
21. Mashaei, M.; Lennartson, B. Energy Reduction in a Pallet-Constrained Flow Shop through On-Off Control of Idle Machines. *IEEE Trans. Autom. Sci. Eng.* **2013**, *10*, 45–56. [CrossRef]
22. Jia, Z.; Zhang, L.; Arinez, J.; Xiao, G. Performance Analysis for Serial Production Lines with Bernoulli Machines and Real-Time Wip Based Machine Switch-On/Off control. *Int. J. Prod. Res.* **2016**, *54*, 1–17. [CrossRef]
23. Su, H.; Frigerio, N.; Matta, A. Energy Saving Opportunities and Value of Information: A Trade-off in a Production Line. *Procedia CIRP* **2016**, *48*, 301–306. [CrossRef]
24. Renna, P. Energy Saving by Switch-Off Policy in a Pull-Controlled Production Line. *Sustain. Prod. Consum.* **2018**, *16*, 25–32. [CrossRef]
25. Duque, E.T.; Fei, Z.C.; Wang, J.F.; Li, S.Q.; Li, Y.F. Energy Consumption Control of One Machine Manufacturing System with Stochastic Arrivals Based on Fuzzy Logic. In Proceedings of the 2018 IEEE International Conference on Industrial Engineering and Engineering Management (IEEM), Bangkok, Thailand, 16–19 December 2018; pp. 1503–1507. [CrossRef]
26. Wang, J.; Xue, J.; Duque, E.T.; Li, S.; Chang, Q. Fuzzy Decision of Machine Switch On-Off for Energy Efficient Operation of Manufacturing System. In Proceedings of the 2017 13th IEEE Conference on Automation Science and Engineering (CASE), Xi'an, China, 20–23 August 2017. [CrossRef]
27. Wang, J.; Fei, Z.; Chang, Q.; Li, S.; Fu, Y. Multi-State Decision of Unreliable Machines for Energy-Efficient Production Considering Work-in-Process Inventory. *Int. J. Adv. Manuf. Technol.* **2019**, *102*, 1009–1021. [CrossRef]
28. Marzano, L.; Frigerio, N.; Matta, A. Energy Efficient State Control of Machine Tools: A Time-Based Dynamic Control Policy. In Proceedings of the 2019 IEEE 15th International Conference on Automation Science and Engineering (CASE), Vancouver, BC, Canada, 22–26 August 2019; pp. 596–601. [CrossRef]
29. Otto, A.; Otto, C.; Scholl, A. Systematic Data Generation and Test Design for Solution Algorithms on the Example of SALBPGen for Assembly Line Balancing. *Eur. J. Oper. Res.* **2013**, *228*, 33–45. [CrossRef]

Article

In-Line Target Production for Laser IFE

Irina Aleksandrova, Eugeniy Koshelev and Elena Koresheva *

P.N. Lebedev Physical Institute of the Russian Academy of Sciences, 119991 Moscow, Russia; ivaaleks@gmail.com (I.A.); 2814rosti@mail.ru (E.K.)

* Correspondence: elena.koresheva@gmail.com; Tel.: +7-916-159-3641

Received: 25 December 2019; Accepted: 15 January 2020; Published: 18 January 2020

Featured Application: Fuel target supply to high repetition rate laser facilities including inertial confinement fusion power plant.

Abstract: The paper presents the results of mathematical and experimental modeling of in-line production of inertial fusion energy (IFE) targets of a reactor-scaled design. The technical approach is the free-standing target (FST) layering method in line-moving spherical shells. This includes each step of the fabrication and injection processes in the FST transmission line (FST-TL) considered as a potential solution of the problem of mass target manufacturing. Finely, we discuss the development strategy of the FST-TL creation seeking to develop commercial power production based on laser IFE.

Keywords: inertial fusion energy; free-standing target layering; FST transmission line

---

1. Introduction

In the frame of International Atomic Energy Agency (IAEA) project (IAEA CRP F13016), one of the specific objectives was to define options for fuel target material choice and mass manufacturing methods, requirements, development pathways and potential solutions [1]. In this paper we identify the principle challenges associated with progression from single shot to high repetition rate operation with regard to manufacturing methods of a cryogenic fuel target (everywhere further—target) for laser IFE.

A fusion power plant will consume as many as one million targets per day. Therefore, the free-standing target transmission line (FST-TL) is an integral part of any IFE reactor. Hereupon, methodologies of the fabrication and injection processes must be applicable to mass-production layering at low cost.

To meet the above requirements, the LPI has proposed to use line-moving and free-standing targets to develop a scientific and technological base for repeatable target supply at the laser focus [2–5]. Precisely moving targets co-operate all production steps in the FST-TL that is considered as a potential solution of mass-production layering and noncontact target delivery in pulsed, repetitively cycled IFE systems. This approach includes also the development of nano-layering technologies, which supports the fuel layer survivability under target injection and transport through the reactor chamber (everywhere further—FST layering method) [6–8].

The target production area includes high-precision technologies for target fabrication, injection and tracking. These technologies are currently an important research stage in the leading laboratories of the world [9–18]. Target compression and ignition is regarded as a major challenge to the IFE community. In addition, to maintain an acceptable tritium inventory, it is needed the development of reasonably fast layering techniques. An overview of various fuel layering methods is given in Reference [2,19].

A traditional approach is based on the following concept: targets for laser experiments are filled by permeation, and a uniform D–T ice layer (molecular composition: 25% of D_2, 50% of DT molecules, and 25% of T_2) is formed by "beta layering" method [19–21]. This method involves crystallization from a single seed crystal in the fixed target. The process takes place in the anisotropic hydrogen isotopes with extremely slow cooling (q ~ $3 \cdot 10^{-5}$ K/s) and precise temperature control (<100 μK) for obtaining cryogenic layers like a single crystal. Long-run beta-layering process at very strict isothermal conditions requires ~24 h for one attempt. But routine practice is between 1 and 4 attempts, or even 6 attempts [20,21] which generally require several days or a week. The beta-layering method can form a spherical fuel layer in the uniform thermal environment, but the ability of the D–T targets to withstand acceleration during injection is a key issue in terms of surface perturbation growth in the anisotropic layers under thermal and mechanical loads. Such layers also do not contribute to avoiding the Rayleigh-Taylor instabilities during implosion that decreases the odds for achieving ignition [22]. Moreover, this method is not efficient for repetitively cycled schedule of the target fabrication and injection because the target must be precisely located in the reaction chamber under precise temperature control. We emphasize that even in the single-shot laser experiments an important role for achieving ignition (apart from anisotropic layers) can play some other factors, such as effect of the target support on ice-layer quality, cryogenic shroud retraction, vibration control, target alignment, etc. As noted in Reference [9,13], the beta-layering targets currently cost thousands of dollars apiece; thereby, over the next several years, the focus of the target layering should be on the isotropic fuel structure formation within moving targets to meet the requirements of implosion physics. This places significant onus on the IFE laboratories in the area of new layering methods development.

At the Lebedev Physical Institute (LPI) the researches focus on the extension of the FST layering method (cooling rates q = 1 − 50 K/s) on large cryogenic targets to form isotropic ultrafine solid fuel. Our philosophy in conceptualizing the in-line target production is based on the FST layering method to generate a dynamical symmetrization of the liquid fuel within moving targets (instead of solid fuel redistribution inherent in the traditional approach for motionless targets [19]). During FST layering, two processes are mostly responsible for maintaining a fast fuel layer formation (Figure 1):

- Firstly, the target rotation when it rolls down along the layering channel (LC) (n-fold-spirals at n = 1, 2, 3) results in a liquid layer symmetrization;
- Secondly, the heat-transport outside the target via the conduction through a small contact area between the shell wall and the LC wall. The spiral LC is a special insert into the cryostat, and it is cooled outside by helium. A contact spot moving along the outer surface of the rotating target results in a liquid layer freezing.

For all these reasons, the FST layering method is a promising candidate for development of the FST-TL at a high repetitively cycled capability intended for mass manufacturing of low cost IFE targets. As a result of long-term research effort, the LPI gained a unique experience in the development of the FST layering module (FST-LM) for target fabrication with an isotropic ultrafine fuel layer inside polymer (CH) shells of 1-to-2-mm diameter. Below, we discuss the FST-LM development of the next-generation for targets over 2 mm in diameter.

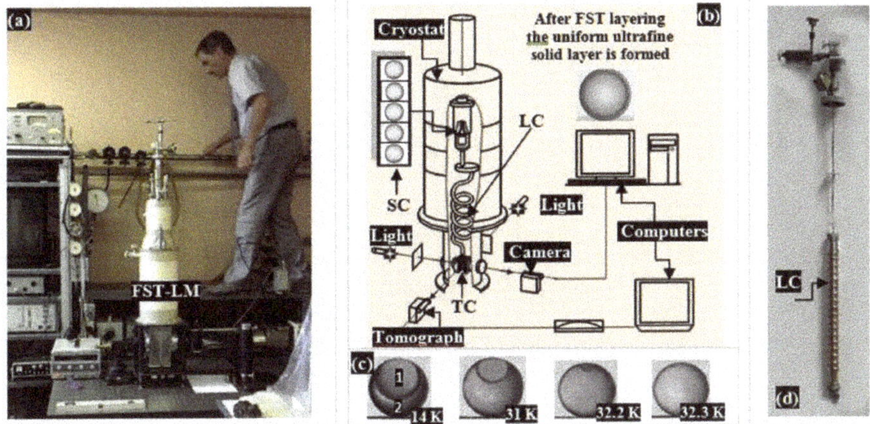

Figure 1. Free-standing target (FST)-layering setup. In (**a**): general view; in (**b**): FST-layering module (LM)—target transport is injection between the basic units: shell container (SC)—layering channel (LC)—test chamber (TC)); in (**c**): (vapor (1) – liquid (2)) interface behavior in 1-mm CH -shell (P = 765 atm of H2 at 300 K); in (**d**): LC as a part of the insert into the cryostat.

2. Modeling Results

2.1. FST-Layering Method for Classical High Gain Targets

Among the reactor-scale targets is a classical high gain target (CHGT-1) with the following parameters [23]: it is a 4-mm diameter CH shell with a 45-μm thick wall, the cryogenic layer thickness is 200 μm. For the CHGT-1, in our previous work we calculated the value of τ_{form} for different initial target temperatures T_{in} [6] (Table 1).

Table 1. FST layering time for classical high gain target (CHGT-1).

	CHGT-1 Design	τ_{form}	
		T_{in} = 37 K	T_{in} = 27 K
1	CH shell without overcoat	227.5 c	149.0 c
2	Gold-coated CH shell	13.65 c	8.94 c

Another example of a classical high gain target (CHGT-2) was proposed in Reference [24]. The target consists of four parts: a gold-coated CH shell, a D–T filled CH foam ablator, a layer of pure solid D–T fuel, and a D–T vapor cavity. We use this CHGT-2 to examine issues affecting the possibility of its fabrication by the FST layering method taking into account that it has a very thin CH shell (1 μm) compared with CHGT-1 (45 μm).

The CHGT-2 specifications are given in Table 2, where D_i and D_0 are the inner and the outer diameters of each layer, and Δ_l and ρ_l are the layer thickness and density, respectively. A thin CH layer has a gold overcoat (0.03 μm) to withstand the thermal chamber environment.

Table 2. CHGT-2 specifications.

Material	D_i (μm)	Δ_l (μm)	D_0 (μm)	ρ_l (g/cm^3)
D−T (vapor)	0	3000	3000	$0.3{\cdot}10^{-3}$
D−T (solid)	3000	190	3380	0.25
CH $(DT)_{64}$	3380	261	3902	0.25775
CH layer	3902	1	3904	1.07

The program on CHGT-2 modeling includes the computation of the FST layering time and the LC specifications, such as: number of spirals, inclination angle of the spiral, total length of the spiral, the spiral diameter, and number of turns. The obtained results will allow one to calculate the baseline parameters necessary at designing of the FST-LM as a means of high-gain direct-drive target production.

For a better understanding of the time-integral performance (TIP) criterion under the FST-LM operation, we emphasize that the LC must have a well-defined geometry to satisfy the condition:

$$\text{Target is formed in the LC if } \tau_{form} \leq \tau_{res}, \quad (1)$$

where τ_{form} is the layering time, and τ_{res} is the time of target residence in the LC. Since a liquid sagged fuel is the initial state before the FST layering (Figure 1c), then for dynamical fuel symmetrization in a batch of rolling targets the time of liquid phase existence, τ_{liquid}, is a key parameter and must be sufficient for layer symmetrization. This depends on the temperature T_{in} of the target entry into the LC and on the temperature distribution along the LC, i.e., the temperature profiling procedure can play an important role for successful FST layering.

Generally, one can view the target motion in the following rolling conditions:

- Target slides on the LC surface (no rotation: sliding and only sliding or pure S&S-mode);
- Target combines rolling with sliding (rolling with sliding or mixed R&S-mode);
- Target rolls on the LC surface without sliding (rolling and only rolling or pure R&R-mode).

A key problem of our investigations is that it is necessary to realize only the R&R-mode to avoid the outer shell roughening and to achieve the fuel layer uniformity. Therefore, the TIP criterion can be written in the following type (τ_{rol} is the time of pure target rolling):

$$\tau_{form} \leq \tau_{res} = \tau_{rol}. \quad (2)$$

Thus, determination of the rolling conditions is one of the main problem, which influences the choice of the FST-LM operation including simplifying the physics design and modifying the specifications. First of all we should to determine the spiral angles α for realizing the pure target rolling (R&R-mode).

Consider a cylindrical spiral of radius R_0. In each point of the spiral the tangent makes a constant angle α with the horizontal. Let a spherical target of radius R and mass m is moving with a velocity V (t) along the spiral LC. Then the system of equations in accordance with Newton's second law of motion has the following form:

$$dV/dt = g\sin\alpha - F_f/m, \quad (3)$$

$$(J/R)d\omega/dt = F_f - k_r, \quad (4)$$

where g is the free fall acceleration, J and ω are the moment of the target inertia and the angular velocity of its rotation, F_f is the friction force, N is the reaction of support from the LC wall, and k_r is the rolling friction. If the target rolls without sliding the linear and angular velocities are related by equation of rolling

$$V = \omega R. \quad (5)$$

Excluding the friction force from Equation (3), we obtain the rolling equation:

$$(1+\zeta)dV/dt = g\sin\alpha - k_rN/m,\ \zeta = J/(mR^2). \quad (6)$$

Besides, there is the conditional relation that should be maintained for the accelerated target to be in pure rolling:

$$F_f \le F_{max} = k_sN,\ N = m\cos\alpha\{g^2 + (V^4\cos^2\alpha)/R^2\}^{0.5}, \quad (7)$$

where k_s is the sliding friction. The second component under the root is the square of the centripetal acceleration. Solving together Equations (6) and (7), we obtain the working range $\Delta\alpha$, which is given by two inequalities:

$$k_r = tg\alpha_{min} < tg\alpha < tg\alpha_{max} = \{(1+\zeta)k_s - k_r\}/\zeta. \quad (8)$$

The next step is modeling of the CHGT-2 layering. We used a simulation code based on solving the Stephen's problem [25,26]. Using the data in Table 2 and the data on the fuel heat capacity and conductivity [19], we got the following values:

— $T_{in} = T_s \sim 35$ K: $\tau_{form} = 22.45$ s for D_2 fuel and $\tau_{form} = 28.52$ s for D–T fuel,
— $T_{in} = T_d \sim 28$ K: $\tau_{form} = 12.05$ s for D_2 fuel and $\tau_{form} = 14.25$ s for D–T fuel,

where T_s is the temperature of fuel separation into the liquid and gaseous phases, and T_d is the depressurization temperature at which the excess gas is removed from the SC. Note that depending on the shell strength T_d can be $\sim T_s$.

The final step is the computation a set of the optimization parameters related to the LC geometry to maintain the process of the CHGT-2 fabrication by the FST layering method. Using (8), we have found that the CHGT-2 can be fabricated in the FST-LM by using a double-spiral LC that is confirmed by the experimental modeling with pilot LC mockups. The obtained results are presented below.

1. Double-spiral LC:

- Specifications (baseline design: Spiral 1 + Spiral 2): angle of each spiral – $\alpha = 11.50$, radius of each spiral – 21 mm, height of each spiral – 450 mm, length of each spiral – 2.261 m, total spiral length – 4.52 m, total time of target rolling – 23.49 s;
- $T_{in} \sim T_d \rightarrow$ the double-spiral LC specifications are those at which the TIP criterion (2) is valid for both D2 and D–T fuel;
- $T_{in} \sim T_s \rightarrow$ the TIP criterion is valid for D_2 fuel;
- $T_{in} \sim T_s \rightarrow$ the TIP criterion is not valid for D–T fuel.

Nevertheless, the double-spiral LC can work in the latter case, as well, because the length of Spiral 2 can be extended on ~2 m to meet the TIP criterion (Figure 2). Generally, this approach is workable for any spiral LC because several interchangeable spirals with different parameters can be joined to a baseline LC that depends on the experimental goals.

Figure 2. Test mockup of a double-spiral LC with an extra working spiral having a different inclination angle.

For the CHGT-2 we have considered a three-fold-spiral LC, as well. The obtained results are presented below.

2. Tree-fold-spiral LC:

- Specifications #1 (baseline design: Spiral 1 + Spiral 2 + Spiral 3): angle of each spiral – $\alpha = 16.70$, radius of each spiral – 21 mm, height of each spiral – 88 cm, length of each spiral – 3.066 m, total spiral length – 9.187 m, total time of target rolling – 33.29 s;
- Specifications #2 (Specifications #1 + Spiral 4). Spiral 4: angle – $\alpha = 30$, radius – 21 mm, height – 10.8 cm, length – 2.070 m, total length of Spiral 3 and Spiral 4 – 5.136 m. Other words, this 3-fold LC is designed in a special configuration with an extra short Spiral 4 (combined LC).

The modeling has shown that, in ~5 s after a start, the target motion is carried out with a constant velocity $V_{max} = 0.3$ m/s. The total rolling time for Specifications #1 is 33.29 s. Thus, we can realize the rolling conditions for the CHGT-2 and satisfy the TIP criterion in the case of 3-fold-spiral LC, even with a certain time margin.

Note also that the proposed 3-fold-spiral LC (Specifications #1) can have an extra spiral (Specifications #2) so that two spirals "Spiral 3 + Spiral 4" make a combined LC in which the Spiral 4 parameters differ considerably from those of Spiral 3. Such combined LC consists of these spirals assembled one after another: "acceleration Spiral 3 + deceleration Spiral 4" for controlling the target velocity at the output. For example, in our case, it will take no more than 1.5 s at $\alpha = 30$ to zero the target velocity. The obtained results are summarized in Table 3 in which τ_{2rol} and τ_{3rol} are the rolling times for the double-spiral LC and for the tree-fold-spiral LC, respectively.

Table 3. FST layering for both D_2 and D–T.

D_2 Fuel			
Calculation			Experiment
T_{in}	τ_{form}	τ_{2rol}	τ_{2rol}
35.0 K 27.5 K	22.45 s 12.05 s	23.49 s	23.5 s (min)
D–T Fuel			
Calculation			**Experiment**
T_{in}	τ_{form}	τ_{3rol}	τ_{3rol}
37.5 K 28.0 K	28.52 s 14.25 s	33.29 s 34.79 s	35 s (min)

A few comments should be made regarding the gold-coated CH shell. In this case for both D_2 and D–T fuel the layering time $\tau_{form} < 0.5$ s. This means that in order to have a uniform layer, a temperature profiling along the LC becomes a necessary condition for increase in time of τ_{rol} at temperatures more than the triple point one. In this case T_{in} can be ~21 K as the hydrogen isotope vapor pressures near the triple point determine the minimum operating pressures (do not exceed 0.2 atm [19]) that allows one to consider an injection filling procedure. Filling of the CH shells with a cryogenic liquid fuel is suitable for the FST layering method because the fuel in the shell directly before the layering has a two-phase state "Liquid + Vapor" (Figure 1c).

Thus, in the IFE, the FST layering is a credible pathway to a reliable, consistent, and economical target supply. Currently, our studies on the creation of FST-LM lie in the field of compatibility of its work with a noncontact method for delivering targets to the reactor.

2.2. Noncontact Accelerating of a Target

Creation of a delivery system based on noncontact positioning and transport of IFE targets represents one of the major tasks in the IFE research program. The purpose is to maintain a fuel layer quality during the target acceleration and injection at the laser focus. The ability of the target to withstand acceleration during its transport and injection is a key issue in the design of the target delivery system. The stringent requirements to the delivery process are as follows [13,27]: targets should have a temperature of not higher than 18.3 K; the overloads during its acceleration in the injector can be a = 500–1000g (but there is a desired acceleration limit a <500g in order to significantly reduce risks of the target damage due to mechanical overloads arising in the process of its acceleration; targets must be accelerated to high injection velocities (200–400 m/s) to withstand the reaction chamber environment. Therefore, the target acceleration is carried out in a special capsule—target carrier or sabot, which transmits a momentum of motion to the target.

The LPI program includes much development work on creation of different designs of the hybrid accelerators for IFE target transport with levitation. One of the main directions is an electromagnetic accelerator (EM-AC) + PMG-System, where PMG is the permanent magnet guideway. The operational principle is based on a quantum levitation of type-II, high temperature superconductors (HTSC) in the magnetic field. The target temperature (T = 18.3 K) excludes the use of type-I superconductors as they have the critical temperature T_c < 10 K, i.e., their heating above 10 K destroys their superconductivity. At the current research stage, the concept development of a hybrid accelerator «EM-AC + PMG» is complete and proof-of-principle experiments in the mutually normal magnetic fields are carried out. The proposed accelerator is a combination of the acceleration system (field coils generating the traveling magnetic waves) and the levitation system (PMG including a magnetic track). The experimental setup is shown in Figure 3. The PMG consists of 6 rectangular permanent magnets, and two in the middle were covered with a ferromagnetic plate. Along the magnetic track (or acceleration length) the magnetic field is constant (B = 0.33 T onto the permanent magnet surface), which allows the HTSC-Sabot to move with no energy loss. Normal to the acceleration length, the magnet poles are aligned anti-parallel to each other N-S-N (Figure 3a) that produces a considerably strong gradient along the width of the magnetic track. This gradient prevents the motion of the HTSC-Sabots, and they remain located in the transverse direction (Y-axis in the rectangular coordinate system XYZ).

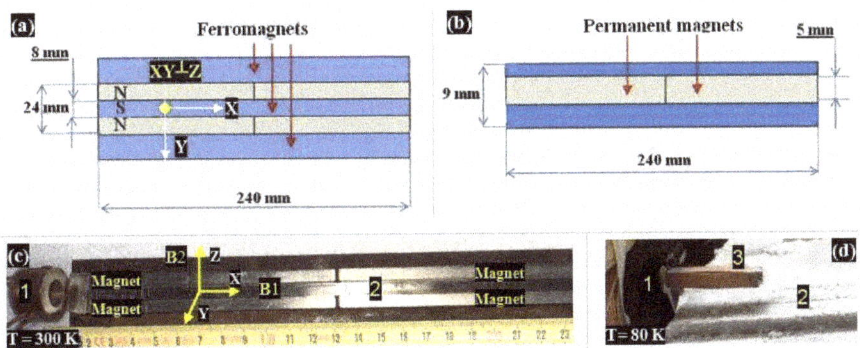

Figure 3. Noncontact acceleration setup. In (**a**,**b**): permanent magnet guideway (PMG)-system dimensions (as a whole, it lies on a ferromagnetic substrate); in (**c**): a top view of the mutual arrangement of the field coil (1) and the PMG-system (2) of two magnets covered with a ferromagnetic plate; in (**d**): a starting location of the high temperature superconductors (HTSC)-Sabot (3), just before the acceleration, with the directions of the levitation force (vector B2 in parallel to the Z-axis) and of the driving force (vector B1 in parallel to the X-axis), as well as along the acceleration length (X-axis).

In our experiments, the HTSC materials are superconducting ceramics based on $YBa_2Cu_3O_{7-x}$ (or Y123) [28] and superconducting tapes of the second generation based on $GdBa_2Cu_3O_{7-x}$ (or Gd123) [29]. These HTSC materials have practically the same values of the critical temperature (T_c) and critical magnetic field (B_c), but they differ from each other in terms of their densities (Table 4) and the magnetic susceptibility (for Y123 ($|\chi| = 2.5 \cdot 10^{-3}$ [28]) and for Gd123 ($|\chi| = 4.5 \cdot 10^{-4}$ [29]). Recall that the phenomenon of superconductivity exists only below the values of B_c and T_c [30].

Table 4. Parameters of the HTSC-materials [28,29].

HTSC Type	Density (g/cm^3)	B_c at 0 K (T)	T_c (K)
Y123	ρ = 4.33	>45 T	91
Gd123	ρ = 3.25	>45 T	92

The experimental setup (Figure 3) operates based on the following effects: the levitation of the HTSC-Sabot is due to the Meissner effect, and the stability of levitation over the PMG system is ensured by vortex pinning [30]. The experiments were made in the mutually normal magnetic fields: the first is **B**1 (from the field coil to move the HTSC-Sabot) directed along the acceleration length, and the second is **B**2 (from the permanent magnets to counteract the gravity) directed normally to the acceleration length (Figure 3c). The PMG-system was optimized in order to achieve simultaneously large levitation forces and stability of the levitation after small perturbations (Figure 3a–c). Figure 3d shows a starting location of the HTSC-Sabot (3) just before the acceleration.

A key feature of the proposed PMG-system is that it has a simple arrangement. It is assembled from several individual magnets that allows it to be extended to any required acceleration length and create multi-stage accelerators. We start with a simple case. Figure 4 demonstrates the work of a one-stage accelerator.

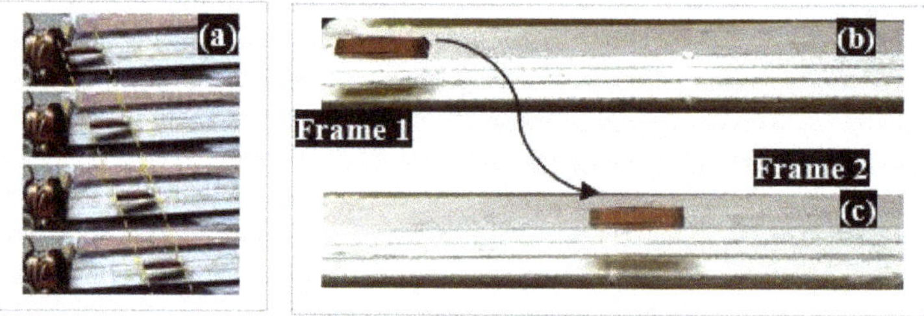

Figure 4. One-stage accelerator: magnetic acceleration of the levitating HTSC-Sabot from Gd123. In (**a**): results of proof-of-principle (POP) experiments: HTSC-sabot (≤30 mm) levitates over the linear PMG-system (~30 cm); in (**b**,**c**): the HTSC-Sabot looks like it is attached to a magnetic track.

The HTSC-Sabot obtains a velocity of 1 m/s (Figure 4a) and keeps it over the whole acceleration length (22.5 cm, see Figure 3c). The vertical and horizontal levitating drift is not observed (Figure 4b,4c). The levitation height is ~3 mm. A set of control experiments has shown that the HTSCs can be successfully used to maintain a friction-free motion of the HTSC-Sabot over the PMG. Besides, they can provide a required stability of the levitation height over the whole acceleration length due to a flux pinning effect [30] inherent in the interaction between a Type-II superconductor and a permanent magnet.

Technologically, the obtained results allow a convenient spacing of the multiple coils (also called a multiple-stages accelerator) and lead to realizing very high velocities of the HTSC-Sabot. In Reference [5], we have reported that, using the MgB_2-superconducting coils [31] as a driving body, it is possible to achieve the required velocity V_{inj} = 200 m/s without exceeding the acceleration limits (a = 400g < 500g), as well as the injector requirement of ~5 m in length. It operates at a very low temperature (~18 K) allowing no heat energy to be passed into the target from the accelerating medium. For future IFE power plants, the injection velocities of V_{inj} > 200 m/s can be easily achieved because the combination of "EM-AC + PMG-System" successfully works as a multiple-stages accelerator.

Therefore, our next step is the development of a multiple-stages accelerator. Below we propose a first version of such accelerator (Figure 5). Superconducting sabot includes several HTSC components (Figure 5a). It comprises not only the accelerating HTSC-coils but also the HTSC-plates for providing its stable levitation along the magnetic track (Figure 5b). During acceleration, the diameter of the barrel exceeds the diameter of the HTSC-Sabot (Figure 5c), allowing it to protect the target against the thermal damage and to reduce the risks from wedging of the sabot in the injector guiding tube.

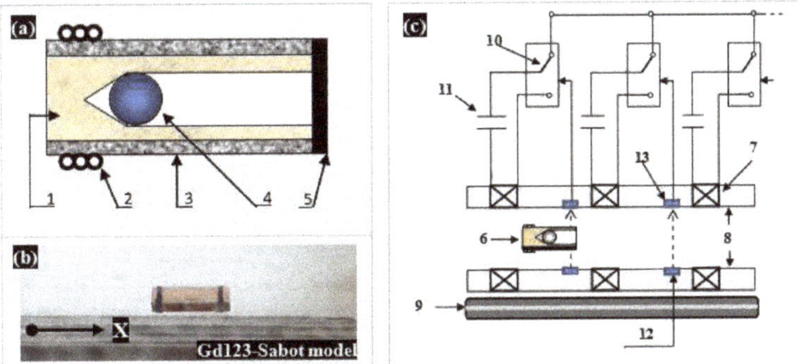

Figure 5. Multiple-stages accelerator for noncontact target transport: (**a**): HTSC-Sabot, which includes 1—polymer matrix, 2—HTSC-coils (MgB_2-driving body), 3—HTSC-plates, 4—target, 5—shield or cover to protect the injected target from a head wind of a residual gas in the reaction chamber; **(b)**: levitation of an HTSC-Sabot mockup (made from the Gd123-tapes) over the linear PMG-system at 80 K (magnetic field is equal to 0.33 T on the permanent magnet surface); (**c**): schematic of the multiple-stages accelerator, which includes 6—HTSC-Sabot, 7—coil, 8—guiding bore, 9—magnetic track, 10—key to run a traveling magnetic wave, 11—capacitor, 12, 13—optocoupler.

Note also that the sequence of field coils loaded on the corresponding capacitance is a line with lumped parameters. From the point of view of the relative position of the accelerating traveling pulse and magnetic dipole (in our case HTSC-Sabot), steady is only the case when an impulse pushes (rather than pulls) a magnetic dipole. This means that the phase (longitudinal) stability is on the front slope of the traveling impulse. Such a technique is called the principle of auto phasing [32] and corresponds to the case of using superconducting materials in the manufacture of the target carriers as they HTSC-Sabot are pushed out from the area of a stronger magnetic field [30].

3. FST-TL Key Elements and Their Functional Description

The free-standing targets are an indispensable requirement for any IFE power plant for continuous target layering and their repeatable injection to the reaction chamber. In other words, targets must remain un-mounted in each production step, and in the IFE experiments, they will be fed directly from the FST-LM to the assembly module with HTSC-Sabots.

3.1. FST Layering Module

Our previous results, as well as the expert analysis carried out in the frame of IAEA CRP 13016 (under the IAEA contract No. 20344 [27]), have shown that the proposed FST-LM (see Figure 1) can meets the requirements for CHGT-2 fabrication.

The construction of the FST-LM is shown in Figure 6.

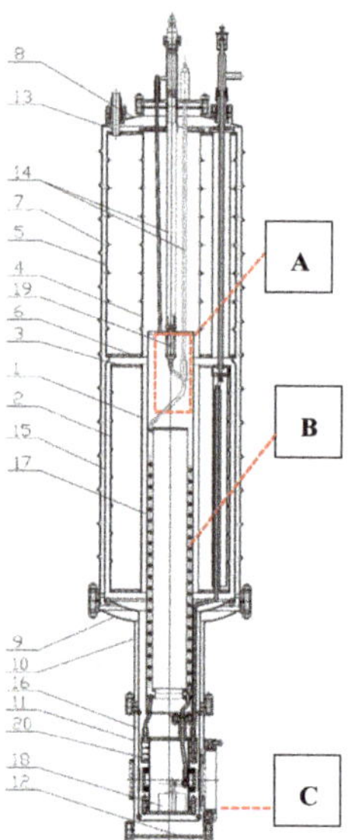

Figure 6. The FST-LM construction: (**A**) Protection-sluice module, which serves to accommodate the shell container (SC); (**B**) Layering channel (LC); (**C**) Test chamber (TC). 1—cryostat housing, 2—nitrogen vessel, 3—thermal shield, 4—helium vessel, 5—well, 6—internal cavities of the insert, 7—liquid-nitrogen transmission line, 8—PSM, 9—place of the SC disposal in PSM, 10—LC, 11—target collector, 12—insert into the helium cryostat well, 13—cryostat neck, 14—traction valve, 15—He-gas output, 16—heat exchanger, and 17—tube space.

The protective-sluice module (Figure 6, item A) serves to accommodate the SC. The LC is made in the form of a tubular spiral (Figure 6, item B), and the outer surface of which is carefully polished and covered with a multilayer screen made of a mylar film to further reduce a degree of the surface blackness.

3.2. FST-Projectile Assembly Module

Before the acceleration and injection stages, the free-standing targets must be mounted in a special module in the form "HTSC-Sabot + Target + Cover" (everywhere further—FST-Projectile). Each step of the assembly process will be operated at significant rates. Several remarks should be made here. In the FST-Projectile, the HTSC-Sabot plays a major role allowing for: (a) to transmit effectively an acceleration pulse to the target, and (b) to protect the target from damage during the acceleration process. An important characteristic of its design is the shape of a target nest. Our study has shown that the shape of the target nest allows one to significantly increase the upper limit of the allowable mechanical overloads on the target and to minimize the injector dimensions. In optimal case the sabot nest should have a conical shape with the angle of 87^0 [4]. A chart of the HTSC-Sabot with the target placed in the conical nest and the cover is shown in Figure 5a.

The FST-projectile assembly module is shown in Figure 7.

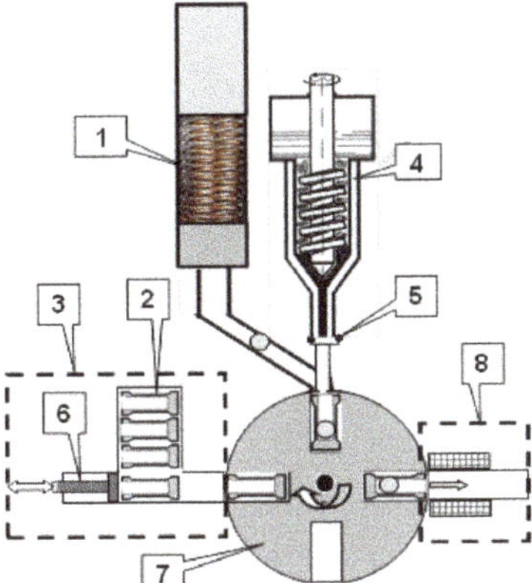

Figure 7. Repeatable assembly of the FST-projectile for safe noncontact delivery of the free-standing targets at the laser focus: 1—FST-LM including a double-spiral LC; 7—target accumulator and system for target transport under gravity to the rotating drum; 3—HTSC-Sabot feeder; 2—a set of the HTSC-Sabots; 3—HTSC-Sabot feeder; 4—extruder of solid D2 cylinder (protective cover production); 5—heater to cut the cover from D2 cylinder; 6—electromagnetic pusher; 7—rotating drum; 8—field coil for the FST-projectile transport to the starting point of the acceleration system.

3.3. Noncontact Delivery System

Our near-term goal is to reduce the acceleration length (~5 m) of the linear accelerator by using a circular PMG of 0.5 m in outer diameter (OD) in the HTSC magnetic levitation (maglev) transport system. This HTSC-maglev is intended for the high target velocity applications, target trajectory correction, and creation of a precise injector.

First, experiments with an acceleration length ~31.5 cm (corresponding to one full turn of a median circle) were made to demonstrate this approach on laboratory scale tests. Figure 8 shows the process of HTSC-Sabot moving at T = 80 K over the circular PMG-system under an electromagnetic driving pulse. The HTSC-Sabot is the same as in the experiments with the linear accelerator (Figure 5b). It is a hollow parallelepiped with dimensions: width 4 mm, height 4 mm, and length 24 mm, which made from the Gd123-tapes with a thickness of 0.3 mm. The HTSC-Sabot mass is 0.2 g. The PMG-system is a neodymium magnet with OD = 10 cm and thickness 5 mm. The internal diameter (ID) is 5 cm. This disc magnet is placed in a ferromagnetic housing open from above (housing thickness 1 mm). Such a configuration allows achieving the desired magnetic field profile over the PMG-system (Figure 8a).

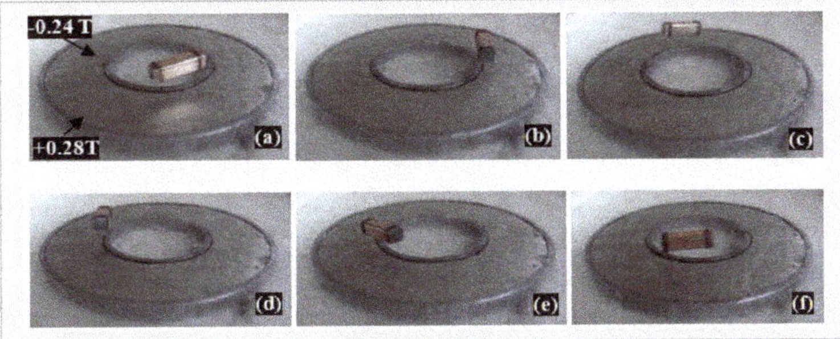

Figure 8. Laboratory scale tests with the PMG-system of a circular type (**a**–**f**—consecutive video frames of the HTSC-Sabot counterclock-wise moving).

3.4. Physical Layout of the FST-TL

Currently, the main problems for target fabrication and injection are as follows:

- There are considered various target designs for direct drive IFE (e.g., compare CHGT-1 and CHGT-2);
- For each target design, low cost methods of high rate target fabrication are needed;
- Targets must survive mechanical and thermal loads during injection;
- Noncontact options for target acceleration and repeatable injection are also needed.

A schematic of the FST-TL of repeatable operation satisfying the listed above requirements is shown in Figure 9.

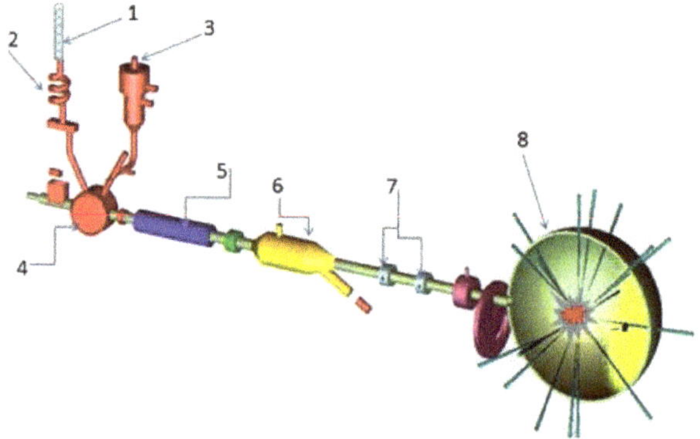

Figure 9. Schematic of the FST- transmission line (TL) operation: 1—SC, 2—FST-LM, 3—extruder for protective cover preparation, 4—FST-Projectile assembly module, 5—noncontact accelerator, 6—FST-Projectile separation module, 7—system for on-line target tracking and characterization, and 8—reaction chamber wall

Note that a scenario of the FST-TL operation in a batch mode has been shown on a reduced scale for the targets under 2 mm in diameter:

- Fuel filling of a batch of 5-to-25 free-standing targets at one time (P_{max} = 1000 atm at 300 K) (Figure 1c);
- Fuel layering within the free-standing and line-moving targets using a single LC (Figure 1d), and the cryogenic layer thickness was up to 100 μm (Figure 1b);
- Levitating HTSC-Sabot acceleration in the mutually normal magnetic fields (Figures 4 and 5b);
- Target injection into the test chamber at a rate of 0.1 Hz at 300 K (Figure 10) and at T = 4.2 K (Figure 11a);
- Target tracking and characterization using the Fourier holography methods.

The last point is not a topic of this report. Therefore we only note the results that are important for a general understanding of the problem of techniques integration into a target production line capable of producing the required amount of the cryogenic targets. As shown in Reference [3], the Fourier holography of the image recognition is a promising way for on-line characterization and tracking of the flying targets. In such a scheme the recognition signal is maximal in case of an exact match between the real and the etalon images; the operation rate is several μs. Achievements in the Fourier target characterization embodied in creation of a simulation code developed as special software "HOLOGRAM". A set of computer experiments have shown that this approach allows to achieve a high accuracy of 0.7 μm in the following operations: (1) recognition of the target imperfections in both low- & high- harmonics, (2) quality control of both a single target and a target batch, and (3) simultaneous control of the injected target quality, its velocity and trajectory.

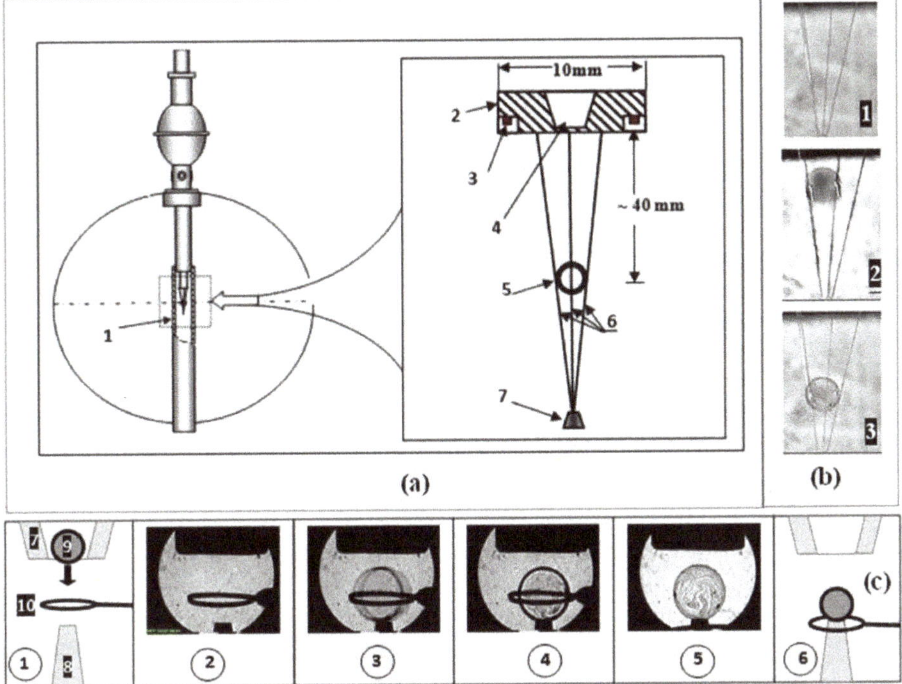

Figure 10. On-holder injection experiments at 300 K (1-6 is the sequence of video frames of the experiment): (a) free target positioning onto the TC bottom; (b) target injection to a special cylindrical cavity imitating a hohlraum-like unit; (c) target injection to a circular PMG-holder.

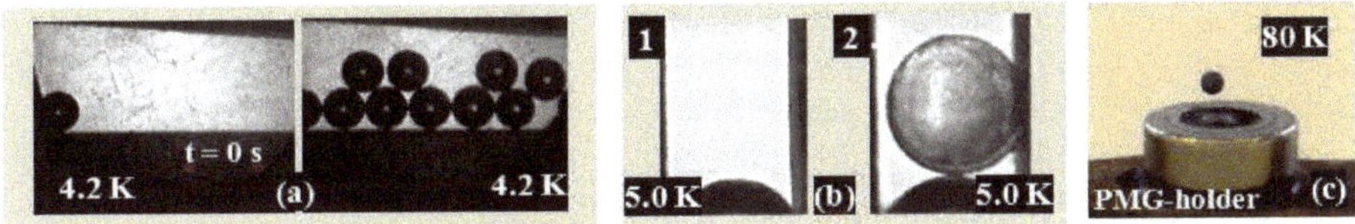

Figure 11. (a–c): A set of the injection experiments at cryogenic temperatures.

4. FST-TL in the Single-Shot Laser Experiments

The FST-TL system shown in Figure 9 can also be used in the laser experiments on a single-shot target irradiation. In this case, it is necessary to capture the injected target on a holder precisely located in advance at the given point. We conducted a series of experiments on the target injection using some different holders: tripod-holder, vacuum suspension on a glass capillary, hohlraum-like unit, and PMG-holder.

In our experiments we have used the gravity injector mounted at the end of the FST-LM. This injector covers a fairly small solid angle (about 15^0) which is no greater than the solid angle of any diagnostic means in the reaction chamber. The target transport by injection is an excellent springboard for overcoming a departure between two concepts: mounted and free-standing targets. Figure 10a shows the relative position of the "injector—holder" couple for a multi-beam laser facility, as well as the results of experimental modeling of the free-standing target assembly with the tripod-holder prearranged at a given point of the test chamber. A final adjustment of the target position relative to the position of the laser focus is carried out using piezoelectric sensors mounted in the tripod base. In Figure 10a, the following designations are accepted: 1—is the retractable cryogenic shroud which must be removed just before the moment of the target irradiation by a laser, 2—output flange; 3—piezoelectric supports made of quartz; 4—nozzle for the target injection from the FST-LM; 5—shell with fuel layer; 6—threads forming a tripod; 7—thread tensioner. The tripod-holder does not require the use of glue for mounting the target, which increases the reliability of operation at cryogenic temperatures and provides a minimum contact area between the target and the threads. The alignment accuracy is +20 μm. The principle of the target alignment is based on the inverse piezoelectric effect which realized by using three rectangular micro-supports made of quartz piezoelectric elements located as shown in Figure 10a (item 3). At the output flange Figure 10a (item 2) there are electrical contacts (not shown in the figure) through which a signal is applied to change the dimensions of these piezoelectric elements. They are small in size and able to work at cryogenic temperatures with a minimal heat loss. The target characterization is based on a backlight shadowgraphy which is a widely used imaging technique. It can provide high contrast object images for a better understanding of an object's shape or size [33]. Specialized software is developed to automatically determine the coordinates of the target and generate commands for correcting its position in real time. Figure 10b shows the details of the injection process on the tripod-holder (1, 2, 3 are the frames before, during, and after the shell injection). The experimental conditions are as follows: CH shell of 2 mm in diameter, the chamber wall temperature is T = 300 K, the pressure inside it is 1 Torr, the holder is the tripod made from three glass fibers (10 μm in diameter). Figure 10c shows the injection of a 300-μm diameter glass shell under gravity from the nozzle of the FST-LM onto the top of a 90-μm diameter capillary prearranged at a given point of the test chamber (here, the target is trapped by the vacuum mount mechanism using this capillary): 1—schematic of the start shell position, 2–5—step-by-step experiment with target injection and its trapping on the top of the capillary, 6—schematic of the final shell position, 7—nozzle, 8—capillary, 9—shell, and 10—restrictive ring used to reliably position the target on the capillary (this ring is removed from the target irradiation area before the laser shot).

Some other options for the target injection and positioning are shown in Figure 11. A unique feature of these experiments is the target injection at cryogenic temperatures. Experimentally, injection can be carrying out in various ways: (1) directly to the TC with a free target positioning onto the TC bottom (Figure 11a), (2) to a special cylindrical cavity imitating a hohlraum-like unit (Figure 11b), and (3) to a circular PMG-holder, which is a promising way for non-contact manipulation and positioning of the cryogenic targets (Figure 11c). In the latter case, for an active target guidance we used an Y123-coated CH shell of 2 mm in diameter (Y123-layer thickness ~10 μm) and a disc magnet with OD = 15 mm, ID = 6 mm, and thickness 5 mm. A ferromagnetic insert is located in the central hole of the magnet. The magnetic field on the periphery of the disc magnet is about 0.3 T.

In the IFE research, these results attract a significant interest due to their potentials for development of the HTSC-maglev suspension technologies and can be used for enhancement of the operating efficiency of the injection process in the single-shot laser experiments.

5. Conclusions

In the direct-drive approach to IFE, the free-standing cryogenic targets of high gain design are required for fueling a laser energy plant. The targets must be delivered to the chamber center with a high accuracy and a high repetition rate. Because fusion reactions must occur approximately ten times per second, the FST-TL becomes an integral part of any fusion reactor. The FST-TL consists of a target production facility, target acceleration, and injection and tracking systems. Currently, target production is in transition from one-of-a-kind to one million of high quality ignition targets per day. Creation of the target delivery system based on noncontact positioning and transport of the cryogenic targets represents one of the major tasks, as well. Therefore, in the frame of IAEA CRP F13016 [1,27], of particular concern is development of a conceptual design of FST-TL including a mass-manufacturing of reactor-scaled targets and their noncontact delivery at the laser focus.

Achieving controlled thermonuclear ignition is one of the goals for IFE, and a target with the isotropic fuel structure is a critical aspect of successfully reaching that goal. This indicates that currently target fabrication technologies for IFE laser experiments are a challenge. A promising candidate is the FST layering method—fast fuel layering inside moving free-standing targets—which is well suited to economical mass production of IFE targets. The cost of the targets produced by the FST layering method may be several orders of magnitude lower than for the traditional target fabrication capabilities and technologies.

In this paper, high-gain direct-drive cryogenic targets using foam to support the D–T fuel were analyzed. These are the shells of ~4 mm in diameter with a wall from compact and porous polymers. The layer thickness is ~200 μm for pure solid fuel and ~250 μm for in-foam solid fuel. A credible path to solve the issue is the FST layering method developed at the LPI. This method is unique and has no analogs in the world. The computation showed that the layering time for such targets was τ_{form} < 23 s for D_2 and τ_{form} < 30 s for D–T fuel, and they can be successfully fabricated by the FST-layering method inside the n-fold LCs at n = 2, 3. So, fast fuel layering is a necessary condition for the producing targets in the massive numbers and for obtaining fuel as isotropic ultrafine layers. Namely, isotropic fuel structure supports the layer quality survivability under the target acceleration, injection, and transport through the reactor chamber. For on-line characterization and tracking of the injected targets it is planned to apply the method of coherent optics based on the Fourier holography.

Additionally, throughout this paper, the principle of innovative HTSC-maglev transport systems are explored for the safe, stable and noncontact target delivery at the laser focus. We address this question both by experimentally demonstrating such system performance and by calculating their parameters. The HTSC levitation is based on the interaction between the magnet and the HTSC. Due to unique characteristics, the HTSC materials have demonstrated tremendous potential for the noncontact target transport.

Any maglev scheme has to address three basic properties: levitation, guidance, and acceleration. The characteristics of the permanent magnets composing the guideway in the PMG-system are very important in terms of levitation force and stability. We found that not always complicated magnet arrangements bring significant improvements with respect to some simpler arrangements that also provide large levitation force. In our study we use two simple magnet arrangements: linear and circular. In the experiments with one-stage linear accelerator, the HTSC-Sabot from the second generation tapes Gd123 gains a speed up to ~1 m/s which remains the same over a 22.5-cm magnetic track (movement time is 0.22 s). The calculations have shown that, using the driving body from MgB_2 superconducting cables as an HTSC-Sabot component, allows reaching the injection velocities of 200 m/s at 5-m-acceleration length without exceeding the established restrictions related to the target acceleration (a = 400 g < 500 g). These velocities can be obtained by using a set of the field coils the number of which is more than 100 pieces (multiple-stages accelerator). Besides, our experiments with strongly pinned superconductors (HTSC-sabot) and permanent magnets in the PMG-system display high stability of the levitation and acceleration processes. This is directly related to the safe operation and design of the whole delivery system.

Recent results obtained at the LPI may help improving the actual design of HTSC-maglev accelerators. Significant reduction of the accelerator dimensions and the number of the field coils can be obtained in a circular accelerator, in which only several field coils are arranged in a circle in the PMG-system. In such geometry, the HTSC-sabot will move along a practically constant circular orbit in pulsed, repetitively cycled mode and will gradually gain the required velocity. The LPI is currently making major efforts to create such accelerator to reach the injection velocities in the range of 200–400 m/s with a constant PMG dimension and a constant number of the field coils. This is a novel way to enhance the performance of target accelerators through HTSC guiding technology.

Thus, a unique scientific, engineering and technological base developed at the LPI allows creating a FST-TL prototype for mass targets production with the ultrafine fuel structure and their noncontact delivery at the laser focus. Such targets for application to high repetition rate laser facilities allow one to test the reactor-scaled technologies and to run a pioneering research of laser direct drive using, for the first time, the isotropic hydrogen fuel in the target compression experiments, including a possibility to apply the FST-TL for single-shot laser experiments for achieving the laboratory-based ignition.

Author Contributions: Conceptualization, I.A. and E.K. (Elena Koresheva); software, I.A.; investigation, E.K. (Eugeniy Koshelev) and E.K. (Elena Koresheva); writing—original draft preparation, I.A.; writing—review and editing, E.K. (Elena Koresheva); project administration, E.K. (Elena Koresheva). All authors have read and agreed to the published version of the manuscript.

Funding: This research was funded by the International Atomic Energy Agency (https://www.iaea.org) under Research Contract No. 20344 and by the Russian Government in the frame of the State Task Program.

Conflicts of Interest: The authors declare no conflict of interest.

References

1. Pathways to Energy from Inertial Fusion: Materials beyond Ignition. In Proceedings of the 3rd IAEA Research Coordination Meeting of the Coordinated Research Project (F13016), Vienna, Austria, 21–23 January 2019; Available online: https://www.iaea.org/events/evt1805417 (accessed on 23 January 2019).
2. Aleksandrova, I.; Koresheva, E. Review on high repetition rate and mass production of the cryogenic targets for laser IFE. *High Power Laser Sci. Eng.* **2017**, *5*, e11. [CrossRef]
3. Aleksandrova, I.; Koresheva, E.; Koshelev, E.; Krokhin, O.; Nikitenko, A.; Osipov, I. Cryogenic hydrogen fuel for controlled inertial confinement fusion (concept of cryogenic target factory based on FST-layering method). *Phys. At. Nucl.* **2017**, *80*, 1227–1248. [CrossRef]
4. Aleksandrova, I.; Koresheva, E.; Koshelev, E. Multiple target protection methods for target delivery at the focus of high repetition rate laser facilities. *Probl. At. Sci. Technol. Ser. Thermonucl. Fusion* **2018**, *41*, 73–85.
5. Aleksandrova, I.; Akunets, A.; Koresheva, E.; Koshelev, E. Contactless HTSC sabot acceleration by mutually normal magnetic fields generated in a PMG system with magnetic travelling wave. In Proceedings of the XLV International Conference on Plasma Physics and Controlled Thermonuclear Fusion, Moscow, Russia, 2–5 April 2018.
6. Aleksandrova, I.; Koresheva, E.; Osipov, I.; Timasheva, T.; Tolokonnikov, S.; Panina, L.; Belolipetskiy, A.; Yaguzinskiy, L. Ultrafine fuel layers for application to ICF/IFE targets. *Fusion Sci. Technol.* **2013**, *63*, 106–119. [CrossRef]
7. Aleksandrova, I.; Bazdenkov, S.; Chtcherbakov, V. Rapid fuel layering inside moving free-standing ICF targets: Physical model and simulation code development. *Laser Part. Beams* **2002**, *20*, 13–21. [CrossRef]
8. Aleksandrova, I.; Bazdenkov, S.; Chtcherbakov, V.; Gromov, A.; Koresheva, E.; Koshelev, E.; Osipov, I.; Yaguzinskiy, L. An efficient method of fuel ice formation in moving free standing ICF/IFE targets. *J. Phys. D Appl. Phys.* **2004**, *37*, 1163–1179. [CrossRef]
9. Goodin, D.; Aleksander, N.B.; Brown, L.C.; Frey, D.; Gallix, R.; Gibson, C.R.; Maxwell, J.L.; Nobile, A.J.; Olson, C.L.; Petzoldt, R.W.; et al. A cost-effective target supply for inertial fusion energy. *Nucl. Fusion* **2004**, *44*, S254–S265. [CrossRef]
10. Canaud, B.; Fortin, X.; Garaude, F.; Meyer, C.; Philippe, F. Progress in direct-drive fusion studies for the Laser Mégajoule. *Laser Part. Beams* **2004**, *22*, 109–114. [CrossRef]
11. Frey, D.; Alexander, N.; Bozek, A.; Goodin, D.; Stemke, R.; Drake, T.; Bitner, D. Mass production methods for fabrication of inertial fusion targets. *Fusion Sci. Technol.* **2007**, *51*, 786–790. [CrossRef]
12. Bousquet, J.; Hung, J.; Goodin, D.; Alexander, N. Advancement in glow discharge polymer coatings for mass production. *Fusion Sci. Technol.* **2009**, *55*, 446–449. [CrossRef]
13. National Research Council of the National Academies. Inertial Fusion Energy Technologies. In *An Assessment of the Prospects for Inertial Fusion Energy*; The National Academies Press: Washington, DC, USA, 2013; Chapter 3; pp. 89–105. ISBN 978-0-309-27081-6.

14. Mori, Y.; Nishimura, Y.; Ishii, K.; Hanayama, R.; Kitagawa, Y.; Sekine, T.; Takeuchi, Y.; Satoh, N.; Kurita, T.; Kato, Y.; et al. 1-Hz Bead-Pellet Injection System for Fusion Reaction Engaged by a Laser HAMA Using Ultra-Intense Counter Beams. *Fusion Sci. Technol.* **2019**, *75*, 36–48. [CrossRef]
15. Mori, Y.; Iwamoto, A.; Ishii, K.; Hanayama, R.; Okihara, S.; Kitagawa, Y.; Nishimura, Y.; Komeda, O.; Hioki, T.; Motohiro, T.; et al. Spherical shell pellet injection system for repetitive laser engagement. *Nucl. Fusion* **2019**, *59*, 096022. [CrossRef]
16. Ren, L.; Shao, P.; Zhao, D.; Zhou, Y.; Cai, Z.; Hua, N.; Jiao, Z.; Xia, L.; Qiao, Z.; Wu, R.; et al. Target Alignment in the Shen-Guang II Upgrade Laser Facility. *High Power Laser Sci. Eng.* **2018**, *6*, e10. [CrossRef]
17. Tsuji, R. Trajectory adjusting system using a magnetic lens for a Pb-coated superconducting IFE target. *Fusion Eng. Des.* **2006**, *81*, 2877–2885. [CrossRef]
18. Kubo, T.; Karino, T.; Kato, H.; Kawata, S. Fuel Pellet Alignment in Heavy-Ion Inertial Fusion Reactor. *IEEE Trans. Plasma Sci.* **2019**, *47*, 2–8. [CrossRef]
19. Kucheyev, S.O.; Hamza, A.V. Condensed hydrogen for thermonuclear fusion. *J. Appl. Phys.* **2010**, *108*, 091101. [CrossRef]
20. Koziozemski, B.J.; Mapoles, E.R.; Sater, J.D.; Chernov, A.A.; Moody, J.D.; Lugten, J.B.; Johnson, M.A. Deuterium-Tritium Fuel Layer Formation for the National Ignition Facility. *Fusion Sci. Technol.* **2011**, *59*, 14–25. [CrossRef]
21. Mapoles, E. Production of hydrogen ice layers for NIF targets. In Proceedings of the 7th International Conference on Inertial Fusion Science and Applications, Bordeaux, France, 12–16 September 2011.
22. Bringa, E.M.; Caro, A.; Victoria, M.; Park, N. Atomistic modeling of wave propagation in nanocristals. *J. Miner. Met. Mater. Soc.* **2005**, *57*, 67–70. [CrossRef]
23. Nakai, S.; Miley, G. *Physics of High Power Laser and Matter Interactions*; Word Scientific Publishing: Singapore, 1992; Volume 1, pp. 87–90.
24. Bodner, S.; Colombant, D.; Schmitt, A.; Gardner, J.; Lehmberg, R.; Obenschain, S. High Gain Target Design for Laser Fusion. *Phys. Plasmas* **2000**, *7*, 2298–2301. [CrossRef]
25. Tikhonov, A.N.; Samarsky, A.A. *Equations of Mathematical Physics*; Nauka: Moscow, Russia, 1977.
26. Samarsky, A.A.; Mikhailov, A.P. *Mathematical Modeling: Ideas, Methods, Examples*; Nauka: Moscow, Russia, 2002.
27. Koresheva, E. FST transmission line for mass manufacturing of IFE targets. In Proceedings of the 1st IAEA RCM of the CRP Pathways to Energy from Inertial Fusion: Materials beyond Ignition, Vienna, Austria, 16–19 February 2011; IAEA: Vienna, Austria, 2016.
28. Golovashkin, A.I.; Ivanenko, O.M.; Leitus, G.I.; Mitsen, K.V.; Karpinskii, O.G.; Shamrai, V.F. Anomalous behavior of the structural parameters of the ceramic YBa2Cu3O7 near the superconducting transition. *J. Exp. Theor. Phys. (JETP) Lett.* **1987**, *46*, 410–412.
29. Lee, S.; Petrykin, V.; Molodyk, A.; Samoilenkov, S.; Kaul, A.; Vavilov, A.; Vysotsky, V.; Fetisov, S. Development and production of second generation high Tc superconducting tapes at SuperOx and first tests of model cables. *Supercond. Sci. Technol.* **2014**, *27*, 044022. [CrossRef]
30. Landau, L.; Lifshitz, E. *Theoretical Physics. Electrodynamics of Continuous Media*, 2nd ed.; Nauka: Moscow, Russia, 1982.
31. Goldacker, W.; Schlachter, S.; Zimmer, S.; Reineret, H. High transport currents in mechanically reinforced MgB2 wires. *Supercond. Sci. Technol.* **2001**, *14*, 783–786. [CrossRef]
32. Dolya, S. Acceleration of magnetic dipoles by a sequence of current-carrying turns. *Tech. Phys.* **2014**, *59*, 1694–1697. [CrossRef]
33. Settles, G.; Hargather, M. A review of recent developments in schlieren and shadowgraph techniques. *Meas. Sci. Technol.* **2017**, *28*, 042001. [CrossRef]

Article

Statistical Analysis and Data Envelopment Analysis to Improve the Efficiency of Manufacturing Process of Electrical Conductors

Marco Antonio Zamora-Antuñano [1,†], Jorge Cruz-Salinas [2], Juvenal Rodríguez-Reséndiz [3,*,†], Carlos Alberto González-Gutiérrez [1], Néstor Méndez-Lozano [1], Wilfrido Jacobo Paredes-García [3], José Antonio Altamirano-Corro [1] and José Alfredo Gaytán-Díaz [4]

1 Engineering Department, Universidad del Valle de Mexico, 76230 Querétaro, Mexico; marco.zamora@uvmnet.edu (M.A.Z.-A.); calberto.gonzalez@uvmnet.edu (C.A.G.-G.); nestor.mendez@uvmnet.edu (N.M.-L.); jaaltami@gmail.com (J.A.A.-C.)

2 Centro de Investigación en Ciencia Aplicada y Tecnología Avanzada, IPN, 76090 Querétaro, Mexico; j.cruz@inteligenciacolectivakm.com

3 Facultad de Ingeniería, Universidad Autónoma de Querétaro, 76010 Querétaro, Mexico; wparedes17@alumnos.uaq.mx

4 Engineering Department, Universidad Politécnica de Santa Rosa Jáuregui, 76220 Querétaro, Mexico; agaytan@upsrj.edu.mx

* Correspondence: juvenal@uaq.edu.mx; Tel.: +52-442-192-12-00

† These authors contributed equally to this work.

Received: 22 July 2019; Accepted: 19 September 2019; Published: 21 September 2019

Abstract: The main focus of this research was to develop an approach using statistical tools and Data envelopment analysis (DEA) to tackling productivity measurements and benchmarking problems in electrical conductor manufacturing environment. In the present work, a tooling efficiency study was carried out with a nozzle used for the manufacture of 23-AWG wires. The efficiency of five types of tooling, four non-Mexican-manufactured types and one Mexican-manufactured type, were compared. Analysis of Variance (ANOVA) and the Tukey test were applied. Six factors were considered that influence of the performance of the tooling during the manufacturing process: productivity, quality, time, machine, operator, and color of the insulating material, but the research work focuses on the efficiency of the tooling die-nozzle. The results demonstrated that two die-nozzle models exhibited the best performance; one of them was the Mexican model, surpassed by a non-Mexican model, the capability process index Cpk = 1.26 manifested a better performance for the 3DND die-nozzle according to the statistical analysis and the tests performed. Subsequently, through a super-efficiency DEA model of inputs-oriented with non-decreasing returns to scale (NDRS). The results obtained in the statistical analysis were corroborated using this technique, its application combined with statistical tools represents an innovation for knowledge in manufacturing processes of electrical conductors. Input data were obtained at a manufacturer of electrical conductors supplier of the automotive sector in the Querétaro City of Mexico.

Keywords: electrical conductor production; statistical analysis; design of experiments; data envelopment analysis

1. Introduction

ANOVA is a statistical technique widely used to evaluate the importance of one or more factors in a product or a process when comparing the means of the response variable at the different levels of the factors. ANOVA analyses require data from populations that follow an approximately normal distribution with equal variances between factor levels. The name "analysis of variance" is based on the approach in which the procedure uses the variances to determine if the means are different. The procedure works by comparing the variance between the means of the groups and the variance within the groups as a way to determine if the groups are all part of a larger population or separate populations with different characteristics. Presently, it is possible to combine other techniques to give greater effectiveness. Over time, the Design of Experiments (DOE) has been applied with great success in all types of processes and products, mainly in the automotive sector. DOE is used as the first approach for optimization of processes, devices, and manufacture of electrical circuits, pinpointing the factors involved in failures in manufacture processes [1], and identification of elements in the manufacture process to improve the reliability [2]. Based on [3], through the DOE, we made improvements in organic processes. In [4], developed the DOE, improved manufacturing tolerances in permanent magnet motors. Taking into consideration [5], applied the DOE in a spot welding process to identify the key parameters of the process, which influence the tensile strength of welded joints. Also, DOE had been employed successfully en R&R analysis [6], surface response analysis [7] and, algorithm analysis [8]. There have been many cases of successful application of the DOE in all types of production processes. The motivation of this study lies in the interest of analyzing the efficiency of tooling, real monitoring, and measurement processes developed with DOE, and identifying the efficiency of the tooling for the manufacture of electrical conductors. Through a combination of techniques of DOE and DEA, it is intended to establish an analysis of the performance of the tooling and validate the efficiency of the die-nozzle 3DND model of Mexican manufacture versus the die-nozzle models of foreign manufacture. The main problem in the manufacture of electrical conductors is the wear on the die-nozzle tooling. It generates frequent changes and causes stoppages in the production lines and implies a high consumption of the die-nozzle. In addition to this, there is an increase in the indirect costs of the process, losses generated by the non-conforming product, waste of materials and contamination. Therefore, the problem in this research work was defined in terms of the durability of the torque die-nozzle because the accelerated wear of this leads to the loss of productivity due to downtime, waste, and contamination [9]. It was developed in a company that manufactures electrical conductors for suppliers of the automotive sector, the efficiency of die-nozzle tooling was analyzed.

1.1. Manufacturing 23-AWG Wire Process and Die-Nozzle

The insulator of the electrical conductors are applied using a plastic extrusion machine whose schematic cut is displayed in Figure 1. A die-nozzle is inserted in the head. The equipment is complemented with a hopper feeding polymer granules and dryer, which are not shown.

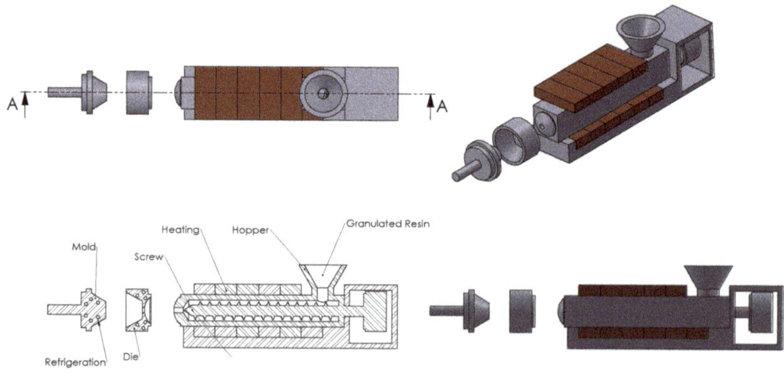

Figure 1. Schematic cutting of the die-nozzle composite head.

In this work, we applied the DOE and DEA to find an optimal solution to the manufacturing problem specifically in an electrical conductor manufacturing industry. The DEA provides a framework for ranking efficient units and facilitates comparison with rankings based on parametric methods [10].

1.2. Data Envelopment Analysis

DEA is an operational research technique focused on measuring and analyzing the efficiency with which goods and services are produced. Its main applications have been in sectors such as finance, health, services, education or to measure the effectiveness of organizational indicators [11].

DEA is a relatively new data-oriented approach for evaluating the performance of a set of peer entities called Decision Making Units (DMUs) which convert multiple inputs into multiple outputs. The definition of a DMU is generic and flexible. Recent years have seen a great variety of applications of DEA for use in evaluating the performance of different kinds of entities engaged in different activities in certain contexts in lots of countries [12].

To differentiate and rank the performance of efficient DMU, [13] proposed the concept of super-efficiency. DEA is capable of transforming a productive situation in which diverse resources generate multiple products in a single index of efficiency. This index is identified with the value that maximizes the quotient between the weighted sum of the outputs and the weighted sum of the inputs of the entity analyzed, to estimate the level of relative efficiency of a company or production unit with respect to the rest of the units that are evaluated simultaneously, making the best use of resources. It provides useful information to improve efficiency by providing tools establishing which cases do not use resources efficiently and which are the recommendations on the use of resources to improve the efficiency of the organization [14].

Analyzing the performance of a set of homogeneous items is very important in the manufacturing processes. Parametric and non-parametric methods are proposed for evaluating the performance of DMUs. DEA is one of the most famous and applicable parametric techniques in this field. It has many applications for interpreting the productivity of complex systems manufacturing [15]. In this project, we made an extensive review covering the literature from 2000 to 2019 and several analyses about DEA. [16] applied the DEA in the design of production lines. In [17], due to advances in data collection systems and analysis tools, data mining (DM) applied DEA and Quality Improvement (QI), in manufacturing. In [18], it is demonstrated how the DEA model might be useful to screen training data so a subset of examples that satisfy the monotony property can be identified. They used real-world health care and software engineering data, managerial monotony assumption, and Artificial Neural Network (ANN) as a forecasting model. [19] used DEA and an analytic hierarchy process (AHP) to measure the performance of ball valve production. In [20], taking into account non-operational factors, the DEA was proposed to measure the technical efficiency, scale efficiency and pure technical efficiency of innovation in Semiconductor industry of China. [21] analyzed the effect of reducing energy savings on the efficiency of innovation and performed the DEA and Tobit regression analysis. This was in agreement with [22], who developed for the Chinese Government a model on the allocation of carbon emission reduction quotas through DEA. [23] established improved manufacturing methods using DEA. In addition, [24] studied recent advances and trends in predictive manufacturing systems in big data environments using DEA. Table 1 illustrates the 23-AWG wire. We intend to analyze the performance of a 3DND die-nozzle model of Mexican manufacturing versus four foreign-made die-nozzle models, and its impact on the variation in the thickness of the insulation to determine the capability and performance of the process of each die-nozzle model, with the range of ± 0.06 mm for a base dimension of 0.58 mm.

Table 1. Features of 23-AWG wire.

AWG	Diameter (mm)	Section mm^2	Number of Turns/cm	kg/km	Resistance (Ω/km)	Capacity (A)
23	0.58	0.26	16.0	2.29	56.4	0.73

2. Methodology

The project was carried out in a manufacturing industry supplier of the automotive sector, in the City of Queretaro, Mexico. The name of the company is not mentioned for reasons of confidentiality. What follows is a census of data was made in the process of extrusion of insulating material in 23-AWG wire. The results were compared vs. the manufacturing standards of each die-nozzle models. Figure 2 depicts the representation of die-nozzle, one manufactured domestically and four imported. Figure 3 shows the tooling code. The data processing was computed by using Minitab v18® [25].

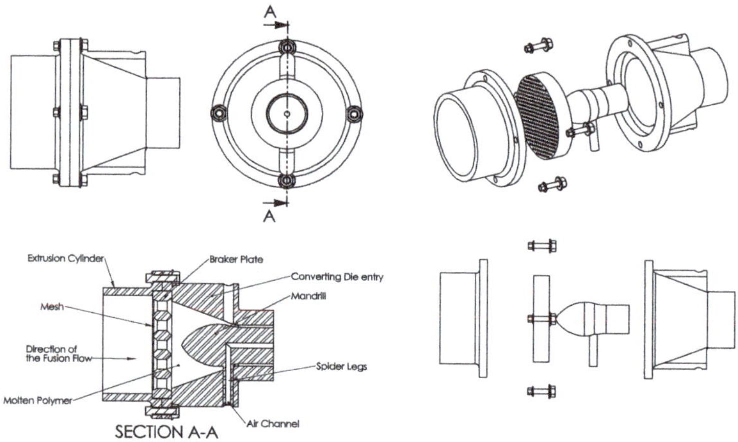

Figure 2. Representation die-nozzle.

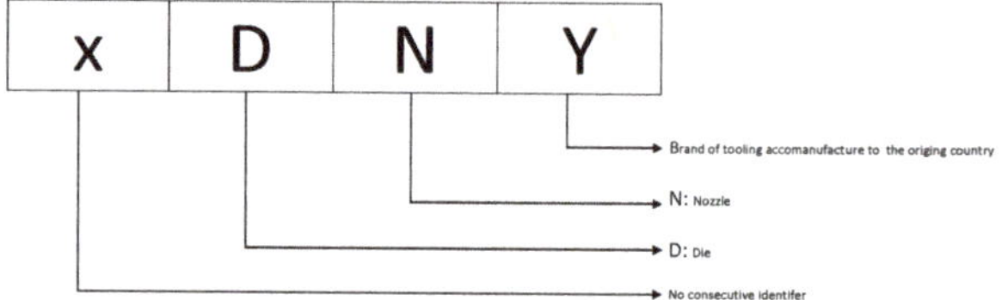

Figure 3. Tooling code for industrial protection purposes.

The manufacture of the 23-AWG wire is ordered in production lines depending on the color of the insulation. The reason for this arrangement is due to the optimization of spaces and the simplification of machine cleaning when a color change of the insulation is made in the manufacturing process [26].

- Line A where the colors black, red, brown and blue are processed.
- Line B with the colors orange, green, yellow and violet.
- Line C that processes the colors white, gray, beige and pink.

There is a total of five work stations in the company, each station is composed of a machine and peripheral equipment, each line produces only four different colors of insulation, the company works four shifts, for the present work the operators of the machines were the same during the whole period of data census. The production of the drivers is subject to market demand, which is the availability of tools of different die-nozzle models. The difference between each die- nozzle pair is defined in the diameter and the material, gibr models of die-nozzle were used, then the manufacturing origin of each model is specified as follows:

- 1DNL: die-nozzle of german manufacturing, diameter: 0.60 mm, material: DIN 420.

- 2DNV: die-nozzle of USA manufacturing, diameter: 0.58 mm, material: ASTM (AISI) 316Ti.
- 3DND: die-nozzle of mexican manufacturing, diameter: 0.59 mm, material: ASTM (AISI) 420.
- 4DNK: die-nozzle of australian manufacturing, diameter: 0.58 mm, material: BSI 316S.
- 5DNS: die-nozzle of japanese manufacturing, diameter: 0.60 mm, material: JIS (SUS 304).

Figure 4 presents the methodology used to apply the DOE and validation with the DEA Model.

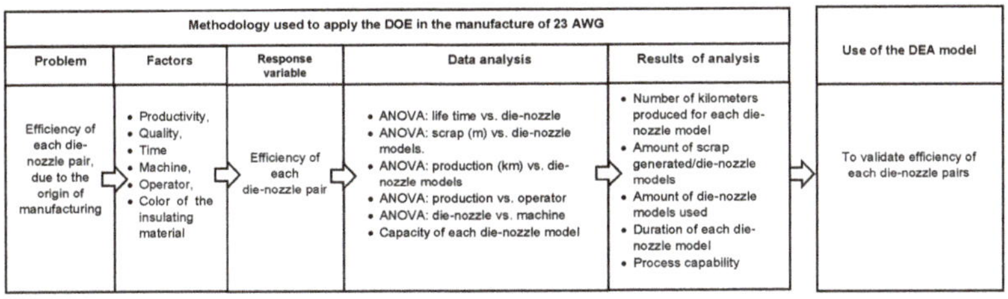

Figure 4. Explain the steps to apply the proposed methodology, data analysis through ANOVA and validation of results with DEA model.

The population that is the object is constituted by the production data for 23-AWG wire: machines, operators, shifts, colors, meters of cable produced, meters of scrap and die-nozzle models that are used for a 23-AWG wire. By convenience sampling, it was determined that the sample would be a proportion of the production from January to June 2018 for the analysis of: Production volume and scrap, variable production vs. die-nozzle quantity, analysis of the production, machine and operator, Key Performance Indicator (KPI) analysis, determination of process capacity indexes, result of analysis and calculation of sigma level [27].

The data available for this investigation correspond to the production history from January to June 2018 (see Table 2), which were registered according to the internal procedures of the company and registered by the production staff directly at the work stations, the data resources correspond to the electrical conductor manufacturing records.

Table 2. Collected data from January to June 2018.

A. Average life duration in hours per month for each die-nozzle model						
Model	**January**	**February**	**March**	**April**	**May**	**June**
1DNL	58	60	57	63	66	80
2DNV	53	48	61	45	56	50
3DND	62	64	56	73	76	69
4DNK	39	46	44	51	37	43
5DNS	49	36	54	40	47	38
B. Average production per unit in meters per month for each model of die and nozzle						
Model	**January**	**February**	**March**	**April**	**May**	**June**
1DNL	719,368	699,942	769,032	783,215	694,142	793,167
2DNV	879,598	759,975	869,432	779,216	865,774	894,302
3DND	689,050	659,920	679,106	746,627	653,511	696,131
4DNK	798,534	789,231	898,535	899,477	775,281	892,387
5DNS	899,506	799,926	877,924	895,685	864,309	894,033
C. Amount of average scrap per unit in meters, generated each month for each model of die and nozzle, used in the extrusion process						
Model	**January**	**February**	**March**	**April**	**May**	**June**
1DNL	7366	7620	7239	8001	8382	101,607
2DNV	9129	8268	10,507	7751	9646	8613
3DND	6402	6608	5782	7537	7847	7124
4DNK	7493	8838	8454	9798	7109	8661
5DNS	10,161	7466	11,198	8295	9747	7880

Due to the manufacture efficiency of the 23-AWG wire can be measured through three indicators: its lifetime, the amount of wire manufactured, and the amount of wire's scrap generated. These indicators were measured during six months for 5 different manufacturing models in order to determine if there was a change at manufacture efficiency of the 23-AWG. However, manufacturing model is not the only factor can affect manufacture efficiency of the 23-AWG wire. Several noisy factors such as: machine, operator, supervisor, shift and, color can alter the manufacturing efficiency [28].

In this way, the following expression can be used to modeling the changes in the manufacturing efficiency:

$$E = \delta_{Model} + \delta_{Machine} + \delta_{Operator} + \delta_{Shift} + \delta_{Color} + \varepsilon \quad (1)$$

where: δ_i is the effect of the factor into the manufacture efficiency and $\in N(0, \sigma^2)$. Nevertheless, as a result of quality control of the company the previous model can be reduced as follows:

$$E = \delta + \delta_{Model} + \varepsilon' \quad (2)$$

where: $\in N(0, \sigma^2)$ with $\sigma < \gamma$. Therefore, the effect of the model can be explored by using a 1-way ANOVA. In Table 2 is presented the measures of the three indicators for every model during the 6-months period and, Figure 5 shows the data analysis workflow. Table 3 presents number of km produced by cable vs. the number of die-nozzles/month, used for each production line.

Table 3. Number of km produced by cable vs. the number of die-nozzles/month, used for each production line.

	Production Line A					
	January	**February**	**March**	**April**	**May**	**June**
Production quantity in km	44,889	40,454	46,736	45,650	43,647	47,956
Number of die-nozzles/month	151	145	152	143	159	138
	Production Line B					
	January	**February**	**March**	**April**	**May**	**June**
Production quantity in km	44,597	39,794	45,476	47,487	43,794	46,433
Number of die-nozzles/month	124	139	134	121	135	145
	Production Line C					
	January	**February**	**March**	**April**	**May**	**June**
Production quantity in km	42,753	38,738	45,488	43,876	41,376	44,289
Number of die-nozzles/month	108	74	111	112	113	88

2.1. DEA Model

The concept of efficiency is related to the economy of resources, efficiency is often defined as the relationship between the results obtained (outputs) and the resources used (inputs). Numerically, an efficiency score can be obtained as a relation between output and input, called technical efficiency. DEA uses linear programming models to determine weights and calculate the efficiency score for each DMUs [29,30]. The only requirements are that the DMUs convert inputs into outputs, both similar and quantifiable. Simultaneously, the multiple inputs and outputs are analyzed in their natural physical units instead of having to convert them to a common denominator [31].

It is assumed that if a die-nozzle DMU, named $die - nozzle_1$ can produce or generate Y_1 output units with X_1 input units, then other die-nozzle must also be able to do the same if they are operated efficiently. Similarly, if $die - nozzle_2$ can produce Y_2 output units with X_2 input units, then the other die-nozzle must also be able to do the same. $Die - nozzle_1$ and $die - nozzle_2$ can be combined to generate a die-nozzle composed of inputs and outputs of them. This virtual die-nozzle is used as a standard of performance for the die-nozzle.

Equation (3) describes the mathematical model of DEA. It allowed to validate the results obtained in the analysis of the performance of the five die-nozzle types. See Section 3.7 Computational results of the DEA model.

Formulation of DEA Model for Die-Nozzle Pairs

$$\begin{aligned} &\min \theta^*, \\ &\text{Subject to} \\ &\sum_{j=1, j\neq 0}^{n} \lambda_j x_{ij} \leq \theta^* x_{io} \text{ for } i = 1, 2, \ldots, m; \\ &\sum_{j=1, j\neq 0}^{n} \lambda_j y_{rj} \leq y_{ro} \text{ for } r = 1, 2, \ldots, s; \\ &\lambda_j \geq 0 \text{ for } j \neq 0; \\ &\sum_{j\neq 0}^{n} \lambda_j \geq 1. \end{aligned} \tag{3}$$

where: DMU_0 represents one of the n DMUs under evaluation, and x_{io} and y_{ro} are the i_{th} input and r_{th} output for DMU_0, respectively. To rate the performance of a particular die-nozzle pair, DEA forms a weighted average of all the observations. The average represents a composite die-nozzle pair. The weights are to be determined so that the composite die-nozzle pair is located on the frontier. It represents *best practice*. The actual die-nozzle pair at hand is then compared with its *best practice*. Denoting the weights to be attached to a particular die-nozzle pair by λ_j. The inputs and outputs for this model are indicated in Table 4. In Table 4, the inputs and outputs of the DMUs are determined, the meaning of these is as follow:

Table 4. DMUs definitions.

Inputs		**Outputs**	
Machine	Time	Productivity	Cpk

- Productivity: meters produced with a specific die-nozzle pair/total meters by period (%).
- Machine: numbers of machine-hours used with a specific die-nozzle pair/total of machine hours/period (%).
- Time: time in hours using a specific die-nozzle pair/total of hours by period (%).
- Cpk: capability process index.

3. Results 23-AWG Production Process

Based on production of 23-AWG, statistical information was given to the information collected for the development of the proposed methodology. The analysis was carried out from January–June 2018. Table 5 presents ANOVA duration vs. die-nozzle model. Table 6 presents the the results of the ANOVA for the three variables: duration, volume of production, and volume of scrap, it can be seen that the value P_{test} is limited to a precision of only 1 thousandth, so any value less than this amount is presented as 0.000 [32]. In this case, it can be observed that there is a significant difference in the average duration of at least two die-nozzle models, with a value of significance α = 0.05 [33]. The results established that the 3DND and 1DNL models are the longest, so they produce more meters of cable than the others.

Table 5. ANOVA: duration vs. die-nozzle model.

Source	DF	SS	MS	F	P
Model	4	2865.5	716.4	15.18	0.000
Error	25	1179.5	47.2		
Total	29	4045.0			

Table 6. ANOVA: scrap (m) vs. die-nozzle models.

Source	DF	SS	MS	F	P
Models	4	19,119,336	4,779,834	4.11	0.011
Error	25	29,095,572	1,163,823		
Total	29	48,214,907			

3.1. Results of Volume vs. Scrap

In Table 6 P_{test} = 0.011 <(α) 0.05, therefore, it can establish that there is a significant difference in the average duration of at least two die-nozzle models.

Table 7A shows the Tukey test, which allowed identifying that the 1DNL and 3DND models have the longest duration and that the 3DND model has an average duration of 2.8 $\pm$ 0.4 days. It was also verified that there is a significant difference in the production of cable and scrap between die-nozzle models. Moreover, Table 7B displays the Tukey test results for each indicator and presents two different groups for production: A(5DNS, 3DND, 4DNK and 2DNV) and B(1DNL), where model group A has higher production that model group B. Finally, Table 7C points to three different groups for scrap: A(5DNS and, 2DNV), AB(4DNK and, 1DNL) and, B(3DND), where model group B presents the lowest scrap generated and two different groups of models for lifetime: A (3DND and 1DNL) and B(2DNV, 5DNS and, 4DNK), where model group A has higher lifetime than model group B [34].

Table 7. Tukey Test.

A. Tukey Test for Duration			
Model	**N**	**Mean**	**Grouping**
3DBD	6	66.67	A
1DBL	6	64	A
2DBV	6	52.17	B
5DBS	6	44	B
4DBK	6	43.33	B
B. Tukey Test Production			
Model	**N**	**Production**	**Grouping**
5DNS	6	871,897.2	A
3DND	6	860,724.2	A
4DNK	6	842,240.8	A
2DNV	6	841,382.8	A
1DNL	6	743,144.3	B
C. Tukey Test for Scrap			
	N	**Scrap**	**Grouping**
5DNS	6	9,124,500	A
2DNV	6	8,985,667	A
4DNK	6	8,392,167	AB
1DNL	6	7,869,167	AB
3DND	6	6,883,333	B

3.2. Results of Data Analysis

The results of the data analysis of the three production lines with the monthly data of the period under study demonstrates the behavior of the operating variables included in the performance evaluation affects the die-nozzle pair, illustrates for this case [35]. Table 2A shows that average life. Table 2B displays the average km of electrical conductor produced with each die-nozzle model. Table 2C shows the scrap generated [36]. To validate the employed model, graphical analysis was run. Figure 5 describes the evidence, which is often an assumption, randomness and, homogeneity of variance were accurate for each ANOVA analysis.

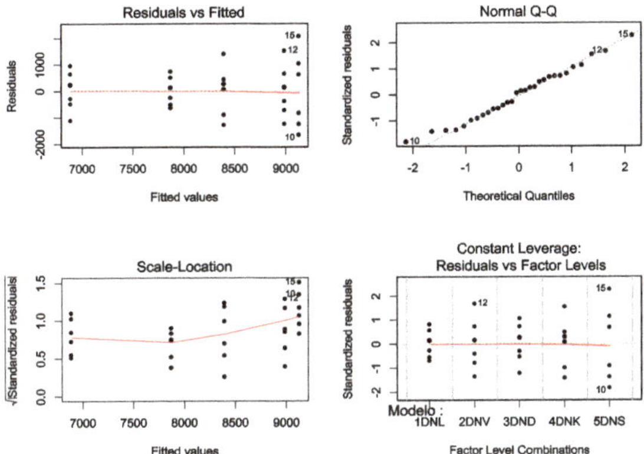

Figure 5. Graphical analysis of residuals of the ANOVA contrast of the model vs. scrap.

In Figures 5 and 6, no significant outliers that affect the normality assumption were detected. A homogeneous variance is maintained in all die-nozzle models.

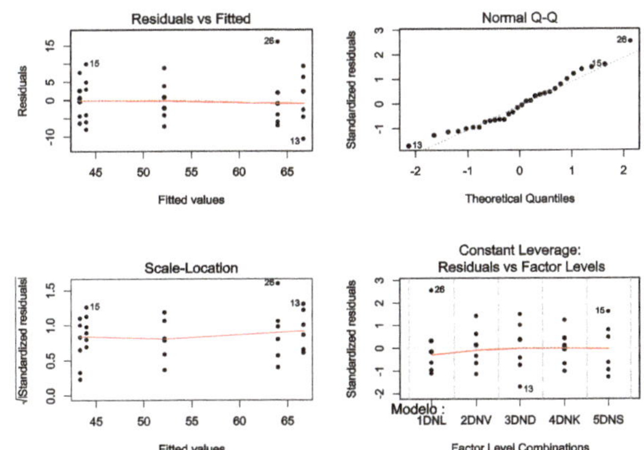

Figure 6. Graphical analysis of the lifetime vs. model.

The results of the statistical analysis applied to the variables of die-nozzles against the production, for each production line, allowed to identify that there is a difference in the production rate between the lines [37]. This is attributed to a dye effect since the production is divided between the lines according to the color of the cable. The dye in the polymers can be obtained using pigments which are inorganic materials, or using dyes which are organic compounds. On the other hand, the tonality is a function of the chemical compound used. In production Line A, the blue dye has a formulation of cobalt aluminate (cobalt oxide and aluminum oxide), which is an abrasive compound, and for the violet which belongs to the group of the production Line B, its formulation is di oxazine (Tetrachloride 1, 4-Benzoquinone), which is a corrosive compound. As can be seen, the composition of the dye formula affects the mechanical elements differently [38].

The results in Table 8, $F_{test} = 28.66$, $F_{table} = 2.76$, and $P_{test} = 0.000 < 0.05 = \alpha$, show that there is a significant difference, with a value of significance $\alpha = 0.05$ in the average duration of at least two die-nozzle models.

Table 8. ANOVA:Production volume vs. die-nozzle models.

Source	DF	SS	MS	F	P
Model	4	9.88E10	2.47E10	28.66	0
Error	25	2.15E10	8.62E8		
Total	29	1.20E11			

3.3. *Results of Analysis Machine, Production and Operator*

Finally, in this case, a behavior similar to that of the Production Lines A and B are observed. The order of origin is not significant. Then, it can be excluded from the model. The coefficients of determination are close to 100%. So, it can be concluded that the production in Line C increases at a rate of 416 ± 28 km per die-nozzle unit. Similarly, the regression analysis is performed for the scrap variable, the results are presented in computational results of the DEA. Figures 5 and 6 are presented these results, which were compared with the proposed methods in [39,40]. Table 9 exemplified the results of ANOVA for machine and operator.

Table 9. Results of ANOVA analysis for the machine and operator variables.

	Parameters through ANOVA.			
Variable	F_{test}	F_{test} **(table value)**	P_{test}	α
Machine	7.39	3.10	0.002	0.05
Operator	129.74	3.10	0.000	0.05

In Table 10, it is observed in the results F_{test} = 129.74 > F_{Table} = 3.10, therefore, in terms of the value P_{test} shows that P_{test} = 0.000 <(α) 0.05, there is a significant difference that at least one operator produces more units/period.

Table 10. ANOVA: production vs. operator.

Source	DF	SS	MS	F	P
Operator	3	22821.2	7607.1	129.74	0
Error	20	1172.6	58.6		
Total	23	23993.8			

3.4. *Key Performance Indicator (KPI) Analysis*

The performance indicators were calculated using the data of Table 3B. Table 11 presents the concentration of results the performance indicators for each die-nozzle models.

Table 11. Concentration of results of the performance indicators for each die-nozzle models.

Model	Performance Indicators						
	Productivity	**Quality**	**Machine**	**Operator**	**Time**	**Color**	**Results**
3DND	0.20	0.13	0.09	0.09	0.27	0.04	81.2%
1DNL	0.12	0.08	0.10	0.10	0.13	0.05	58.7%
2DNV	0.22	0.13	0.08	0.08	0.26	0.04	80.6%
5DNS	0.12	0.08	0.10	0.10	0.14	0.05	59.7%
4DNK	0.10	0.07	0.10	0.10	0.13	0.05	55.1%

3.5. *Determination of Process Capacity Indexes*

The significant values of this analysis are Cp = 1.59, Pp = 1.09, Cpk = 1.26, and PPM = 4726.1 (parts/million) of allowed defects. In Table 12, the results are presented in die-nozzle models, the best results were from the model 3DND the die-nozzle of Mexican manufacturing [41].

Table 12. Summary of the results of the processing capacity of each die-nozzle model.

Model	Parameters of Process Capability					
	Standard Deviation	Cp	Cpk	PPM	PPM General	P_{value}
1DNL	0.025	0.80	0.73	18,931	18,484	0.431
2DNV	0.023	0.86	0.62	32,184	16,086	0.086
3DND	0.001	1.59	1.26	17,540	4726	0.255
4DNK	0.037	0.55	0.52	101,853	69,866	0.736
5DNS	0.025	0.78	0.55	50,796	103,305	0.625

It is essential to emphasize that the natural variability of the process, 6σ, is intrinsic to it and independent of the tolerances assigned. Therefore, if 6σ is less than the range of tolerances to be met, some manufactured products are out of tolerance and non-compliant. If this fact is not taken into account and it is intended to correct based on the readjustment of the process, i.e., modify the centering, the only thing that is achieved is to increase its variability. It can be thought that the process is out of control or unstable in the die of non-Mexican manufacturing because of the variations are caused by somewhat unpredictable behavior. In Tables 13 and 14, it is observed that the 3DND model that shows the production process is control $Cpk > 1$, the other models presents variability, $Cpk < 1$, that does not necessarily imply obtaining products outside of specification. The 3DND model process control allows establishing that the objectives of the project are being met through regular monitoring and measurement of its progress to identify planned variations [42].

Table 13. Summary of sigma level results for each model.

Models	Level Z	PPM	PPM General	Cpk
1DNL	2.08	18,931	18,484	0.73
2DNV	1.85	32,184	16,086	0.62
3DND	3.79	17,540	4726	1.26
4DNK	1.27	101,853	69,866	0.52
5DNS	1.64	50,796.14	103,304.56	0.55

Table 14. DEA model dataset.

	Inputs		Outputs	
Models	Machine	Time (s)	Productivity	Cpk
3DND	0.09	0.27	0.20	1.26
1DNL	0.01	0.13	0.12	0.73
2DNV	0.08	0.26	0.22	0.62
5DNS	0.1	0.14	0.12	0.55
4DNK	0.1	0.13	0.10	0.52

3.6. Result of Analysis and Calculation of Sigma Level

Sigma level indexes of the process are calculated using dimensional data from the die-nozzle. Table 13 indicates the results of the Sigma level analysis for each model die-nozzle. With the analysis of the level of Sigma, the significant results are Z (level of Sigma), PPM and Cpk [36].

It is observed that the 1DNL and 3DND models have the highest levels of sigma, of 2.08 and 3.79 respectively. For these die-nozzle models, the Cpk indexes were 0.73 and 1.26. Those of Cp were 1.59 and 0.80, and the Pp were 1.09 and 0.80 respectively, so these die-nozzle models are the closest ones to the reference established by the norm. For that reason, they are the best capacity and ability of the process, in that order. Models 2DNV, 4DNK, and 5DNS resulted in levels lower than what is specified by the Standard for Electrical Products-Conductors-Wires and cords with PVC insulation 105 °C, for Electronic Uses-Specifications (NMX-J-429-ANCE-1994) [19].

3.7. Computational Results of the DEA Model

To carry out the calculations, the DEA Frontier^{TM} software was used, which is an Excel. DEA Frontier^{TM} uses the Excel solver as the engine to solve DEA models and was used to solve the DEA super efficiency-oriented model to inputs with NDRS [29,30].

4. Discussion and Conclusions

4.1. Discussion

Based on the results obtained by statistical analysis, it can be mentioned that the manufacturing process of 23-AWG wire presented the best performance in terms of the defined variables and results obtained through the DOE. The Cpk of 1.26 allowed to verify that the process with the 3DND model is the most stable compared to the models of this foreign-made die. As for the scrap it generates, the 3DND model was the one with the best performance, as well as the average hours of use, and the second-best performance in terms of total km produced. Being in this last point the model 2DNV the one of better performance. It is proven that through the application of the DOE and process optimization significant results were achieved in continuous improvement for any product or process. In [37] applied the SPC, DOE and Six Sigma in the manufacture of electrical circuits achieving a significant decrease in manufacturing defects. Also many success stories in different scenarios or organizations used the application of statistical tools and techniques improvements are established. Successful cases: In accordance with [6], proposed rapid modeling and optimization of manufacturing processes based on DOE. In [7] is described how processes can be optimized efficiently and how DOE findings may be applied to scale-up. Others [8], used the DOE to improve the manufacturing results in engines. In [9], through DOE it was able to establish critical parameters associated with the processes an indefinite number of success stories in the application of the DOE. The JIT, such as philosophy developed at Toyota and its different methods and systems have served as an example for many organizations to establish continuous improvement, this is a reference for any. In [38], established the methodology allowed the identification of variables that influenced the risk. Davim mentioned that the approach is based on a combination of Techniques of Taguchi and ANOVA [36]. The Statistical Process Control (SPC) is a solution developed to easily collect and analyze data, allowing performance monitoring as well as achieving sustainable improvements in quality which in turn allows increasing the profitability [35]. Based on [36] has proposed a new paradigm in the application of different techniques and improvement tools that imply: Quality, Cost, and Delivery (QCD), and this is to combine techniques and tools already tested as SPC and DOE, with other models in our case the DEA. For [29,30], DEA it is an extremely powerful tool that can assist decision-makers in benchmarking and analyzing complex operational performance issues in manufacturing organizations as well as evaluating processes in banking, retail, franchising, health care, public services, and many other industries. Rau through a combination of SPC and DEA, optimal solutions were introduced in different processes [37].

4.2. Conclusions

ANOVA, with a level of significance $\alpha = 0.05$, was performed on the variables: durability, production, and scrap. These analyses demonstrated that there is a significant difference in the average duration between die-nozzle models. The Tukey test identified that the 1DNL and 3DND models have greater durability. The 3DND model has average durability of 2.8 ± 0.4 days. It was also verified that there is a significant difference between cable and scrap production of all die-nozzle models [29,30].

With the results of the statistical analysis previously described, it is justified that the variables: durability, production, scrap, the color of the insulator, machine, and operator, are considered relevant for the calculation of the performance indicators due to the influence of the die-nozzle pairs, the 3DND model is the best performer in terms of the generated scrap [38,39].

Therefore, the results demonstrated that there is an effect of the model of die-nozzle on the variables. The performance is calculated individually with the method of the cable manufacturer, where the result indicates that the models 1DNL = 81.2% and 3DND = 80.2%, are those with the highest performance indicators [40].

The second phase was carried out applying formal tools of Total Quality Management: Benchmarking, Statistical Process Control and DOE. It was obtained that the 3DND and 1DNL models have a σ level of 3.79 and 2.08. Their indicators Cpk = 1.26, 0.73, Cp = 1.59, 0.80, and Pp = 1.09, 0.80 respectively, so these die-nozzle models are the most approach the references established by the norm, for that reason, they are the best in the capability of the process, in that order [41].

The indicators of performance under the criterion of the analysis of the company evaluate in percentages the factors considering a higher weight to three of the six: productivity, quality and time, obtaining like the result the following order of performance of the die-nozzle models 1DND, 4DNK, 2DNV, and 5DNS. The results of the Statistical Process Control and Six Sigma are based on the evaluation of the variability of the measurements of the finished product and the scrap volume generated, obtaining, as a result, the following order: 3DND, 1DNL, 2DNV, 4DNK, and 5DNS, which It is different because different criteria are used. However, it is observed that the models with the best performance are those that have the highest indicators of processes capability. Therefore, the Manufacturing Company could be used the method of Statistical Control and DOE to evaluate the performance of the die-nozzle pairs [42]. Other authors have applied different statistical techniques and have proposed a routine for evaluated the failure rate and inspection intervals using the first-order reliability method (FORM + Fracture) to alleviate the computational cost of probabilistic defect-propagation analysis. The proposed method is one of the first applying reliability concepts to additive manufacturing (AM) components [1]. It is proven that significant improvements can be generated through the systematic application of the TQM process [2]. With the use and application of statistical analysis, reductions in manufacturing, inspection, materials or methods costs can be achieved [43]. The concept of additive manufacturing [7], will be very useful for the optimization of processes in all types of components.

With the results obtained using the DEA model of Super Efficiency oriented to inputs with non-decreasing returns to scale, the same conclusions were obtained as those obtained through Statistical Process Control and DOE, so that the order of preference of the die-nozzle models is as follows: 3DND, 1DNL, 2DNV, 4DNK, and 5DNS [32,33]. The methods to optimize the design proposed in [1] were very useful for following work in the improvement of tooling and manufacturing processes. By an early control such as that proposed in [2], substantial improvements in the manufacture of electrical conductors and other products can be achieved. In the present project, it is very useful to compare the use of ANOVA in [3,4], to improve the proposed method of analysis. Through optimizing other variables such as those mentioned in [5,6] it is possible to generate substantial improvements that will be taken into account in future work in the optimization of manufacturing processes. It can be concluded that DEA model is a method that can be of great help, and although it will not replace the DOE, if it allows manufacturing companies to perform analysis and process improvements in a faster way.

In the informed literature, there are no application cases to measure the efficiency of the production lines in the manufacturing sector of electrical conductors [28]. With this study, the objective established in the present work was fulfilled.

Author Contributions: Conceptualization, M.A.Z.-A. and J.R.-R.; methodology, J.C.-S.; software, J.A.A.-C.; validation, C.A.G.-G., N.M.-L., and J.R.-R.; formal analysis, J.A.A.-C., J.C.-S., and M.A.Z.-A; investigation and visualization, J.A.G.-D., J.A.A.-C. and J.R.-R.; data curation, J.R.G.-M., W.J.P.-G., and M.A.Z.-A.; writing—original draft preparation; writing—original draft, review, and editing, all the authors.

Funding: This research was partial funded by CONACYT and PRODEP.

Acknowledgments: The authors thank CONACYT for the academic scholarship for CVU No 573233. We appreciate the support of Centro de Investigación en Ciencia Aplicada y Tecnología Avanzada del Instituto Politécnico Nacional (CICATA), Unidad Querétaro. Especially for Master Jorge Cruz Salinas, who was the project leader and main author of the design of the method of validation of die-nozzle tooling, that allowed the project development, and for his dedication to obtaining a Master's Degree in Advanced Technology. We also thank National Research Vice Rectory of UVM-Mexico, for the support provided. The authors recognize in a special way PhD. Ivan Domínguez López and Adrián Luis García García, for the advice provided during the development of the project. The authors appreciate Master Rodrigo Pinto Castillo and César Javier Ortíz Echeverri for his support during the revision of the project.

Conflicts of Interest: The authors declare no conflict of interest.

References

1. Ambühl, S.; Kramer, M.; Sørensen, J. Reliability-based structural optimization of wave energy converters. *Energies* **2014**, *7*, 8178–8200. [CrossRef]
2. Mejia-Parra, D.; Sánchez, J.R.; Ruiz-Salguero, O.; Alonso, M.; Izaguirre, A.; Gil, E.; Palomar, J.; Posada, J. In-Line Dimensional Inspection of Warm-Die Forged Revolution Workpieces Using 3D Mesh Reconstruction. *Appl. Sci.* **2019**, *9*, 1069. [CrossRef]
3. Nguyen, H.T.; Hsu, Q.C. Surface roughness analysis in the hard milling of JIS SKD61 alloy steel. *Appl. Sci.* **2016**, *6*, 172. [CrossRef]
4. Costa, A.; Cappadonna, F.; Fichera, S. A hybrid metaheuristic approach for minimizing the total flow time in a flow shop sequence dependent group scheduling problem. *Algorithms* **2014**, *7*, 376–396. [CrossRef]
5. Ambühl, S.; Kramer, M.; Dalsgaard Sørensen, J. Structural reliability of plain bearings for wave energy converter applications. *Energies* **2016**, *9*, 118. [CrossRef]
6. Arbogast, G.W. A case study: Statistical analysis in a production quality improvement project. *J. Qual. Manag.* **1997**, *2*, 267–277. [CrossRef]
7. Coro, A.; Macareno, L.M.; Aguirrebeitia, J.; López de Lacalle, L.N. A Methodology to Evaluate the Reliability Impact of the Replacement of Welded Components by Additive Manufacturing Spare Parts. *Metals* **2019**, *9*, 932. [CrossRef]
8. Umeda, S.; Nakano, M.; Mizuyama, H.; Hibino, N.; Kiritsis, D.; Von Cieminski, G. *Advances in Production Management Systems: Innovative Production Management towards Sustainable Growth: IFIP WG 5.7 International Conference, APMS 2015*; Springer: Berlin/Heidelberg, Germany, 2015; Volume 459, pp. 7–9. [CrossRef]
9. Boning, D.S.; Mozumder, P. DOE/Opt: A system for design of experiments, response surface modeling, and optimization using process and device simulation. *IEEE Trans. Semicond. Manuf.* **1994**, *7*, 233–244. [CrossRef]
10. Weissman, S.A.; Anderson, N.G. Design of experiments (DoE) and process optimization. A review of recent publications. *Org. Process Res. Dev.* **2014**, *19*, 1605–1633. [CrossRef]
11. Khan, M.A.; Husain, I.; Islam, M.R.; Klass, J.T. Design of experiments to address manufacturing tolerances and process variations influencing cogging torque and back EMF in the mass production of the permanent-magnet synchronous motors. *IEEE Trans. Ind. Appl.* **2013**, *50*, 346–355. [CrossRef]
12. Rowlands, H.; Antony, J. Application of design of experiments to a spot welding process. *Assem. Autom.* **2003**, *23*, 273–279. [CrossRef]
13. Andersen, P.; Petersen, N.C. A procedure for ranking efficient units in data envelopment analysis. *Manag. Sci.* **1993**, *39*, 1261–1264. [CrossRef]
14. Khezrimotlagh, D.; Chen, Y. *Decision Making and Performance Evaluation Using Data Envelopment Analysis*; Springer: Berlin/Heidelberg, Germany, 2018; Volume 269, pp. 217–234. [CrossRef]
15. Koksal, G.; Batmaz, İ.; Testik, M.C. A review of data mining applications for quality improvement in manufacturing industry. *Expert Syst. Appl.* **2011**, *38*, 13448–13467. [CrossRef]
16. Pendharkar, P.C. A data envelopment analysis-based approach for data preprocessing. *IEEE Trans. Knowl. Data Eng.* **2005**, *17*, 1379–1388. [CrossRef]
17. Naghiha, S.; Maddahi, R.; Ebrahimnejad, A. An integrated AHP-DEA methodology for evaluation and ranking of production methods in industrial environments. *Int. J. Ind. Syst. Eng.* **2019**, *31*, 343–359. [CrossRef]

18. Li, H.; He, H.; Shan, J.; Cai, J. Innovation efficiency of semiconductor industry in China: A new framework based on generalized three-stage DEA analysis. *Socio-Econ. Plan. Sci.* **2019**, *66*, 136–148. [CrossRef]
19. Shin, J.; Kim, C.; Yang, H. Does Reduction of Material and Energy Consumption Affect to Innovation Efficiency? The Case of Manufacturing Industry in South Korea. *Energies* **2019**, *12*, 1178. [CrossRef]
20. Emrouznejad, A.; Yang, G.L.; Amin, G.R. A novel inverse DEA model with application to allocate the CO2 emissions quota to different regions in Chinese manufacturing industries. *J. Oper. Res. Soc.* **2019**, *70*, 1079–1090. [CrossRef]
21. Kamarudin, N.F.; Rani, R.M.; Halim, F.A. Improving the Food Manufacturing System by Using Simulation and DEA. In *Proceedings of the Third International Conference on Computing, Mathematics and Statistics (iCMS2017)*; Springer: Singapore, 2019; pp. 555–561. [CrossRef]
22. Liu, G.; Zhou, Z.; Qian, X.; Wu, X.; Pang, W. Multidisciplinary design optimization of a swash-plate axial piston pump. *Appl. Sci.* **2016**, *6*, 399. [CrossRef]
23. Ammons, D.N.; Roenigk, D.J. Benchmarking and interorganizational learning in local government. *J. Public Adm. Res. Theory* **2014**, *25*, 309–335. [CrossRef]
24. Thue, W.A. *Electrical Power Cable Engineering*; CRC Press: Boca Raton, 2012.
25. MINITAB. *Computer Program Manual; Software;* MINITAB: State College, PA, USA, 2014.
26. Yang, Y.; Cao, J. Solving quadratic programming problems by delayed projection neural network. *IEEE Trans. Neural Netw.* **2006**, *17*, 1630–1634. [CrossRef] [PubMed]
27. Tai, Y.T.; Pearn, W.L.; Kao, C.-M. Measuring the Manufacturing Yield for Processes with Multiple Manufacturing Lines. *IEEE Trans. Semicond. Manuf.* **2012**, *25*, 284–290. [CrossRef]
28. Xiao, R.; Liu, B.; Shen, J.; Guo, N.; Yan, W.; Chen, Z. Comparisons of energy management methods for a parallel plug-in hybrid electric vehicle between the convex optimization and dynamic programming. *Appl. Sci.* **2018**, *8*, 218. [CrossRef]
29. Zhu, J. Super-efficiency and DEA sensitivity analysis. *Eur. J. Oper. Res.* **2001**, *129*, 443–455. [CrossRef]
30. Zhu, J. Quantitative models for performance evaluation and benchmarking: data envelopment analysis with spreadsheets. *Int. Ser. Oper. Res. Manag. Sci.* **2014**, *213*, 175–205. [CrossRef]
31. Andreadou, N.; Lucas, A.; Tarantola, S.; Poursanidis, I. Design of Experiments in the Methodology for Interoperability Testing: Evaluating AMI Message Exchange. *Appl. Sci.* **2019**, *9*, 1221. [CrossRef]
32. Han, C.; Sin, I.; Kwon, H.; Park, S. The Role of the Process and Design Variables in Improving the Performance of Heat Exchanger Tube Expansion. *Appl. Sci.* **2018**, *8*, 756. [CrossRef]
33. Lou, P.; Yuan, L.; Hu, J.; Yan, J.; Fu, J. A Comprehensive Assessment Approach to Evaluate the Trustworthiness of Manufacturing Services in Cloud Manufacturing Environment. *IEEE Access* **2018**, *6*, 30819–30828. [CrossRef]
34. Yaste, R.E.; Budrow, R.B. Void-Free Electrical Conductor for Power Cables and Process for Making Same. US Patent 4,319,074, 1982.
35. Realyvásquez-Vargas, A.; Arredondo-Soto, K.; Carrillo-Gutiérrez, T.; Ravelo, G. Applying the Plan-Do-Check-Act (PDCA) Cycle to Reduce the Defects in the Manufacturing Industry. A Case Study. *Appl. Sci.* **2018**, *8*, 2181. [CrossRef]
36. Dyck, T.; Bund, A. Influence of the Bead Geometry and the Tin Layer on the Contact Resistance of Copper Conductors. *IEEE Trans. Compon. Packag. Manuf. Technol.* **2018**, *8*, 1863–1868. [CrossRef]
37. Kilickap, E.; Yardimeden, A.; Çelik, Y.H. Mathematical Modelling and Optimization of Cutting Force, Tool Wear and Surface Roughness by Using Artificial Neural Network and Response Surface Methodology in Milling of Ti-6242S. *Appl. Sci.* **2017**, *7*, 1064. [CrossRef]
38. Sorensen, D.G.H.; Brunoe, T.D.; Nielsen, K. A classification scheme for the processes of the production system. *Procedia CIRP* **2018**, *72*, 609–614. [CrossRef]
39. Yang, C.M.; Lin, K.P.; Chen, K.S. Confidence Interval Based Fuzzy Evaluation Model for an Integrated-Circuit Packaging Molding Process. *Appl. Sci.* **2019**, *9*, 2623. [CrossRef]
40. Godina, R.; Pimentel, C.; Silva, F.; Matias, J.C. Improvement of the Statistical Process Control Certainty in an Automotive Manufacturing Unit. *Procedia Manuf.* **2018**, *17*, 729–736. [CrossRef]
41. Amasaka, K. Applying New JIT, a management technology strategy model at Toyota-Strategic QCD studies with affiliated and non-affiliated suppliers. *Int. J. Prod. Econ.* **2002**, *80*, 135–144. [CrossRef]

42. Rau, H.; Wu, C.H.; Shiang, W.J.; Huang, P.T. A decision support system of statistical process control for printed circuit boards manufacturing. In Proceedings of the 2010 International Conference on Machine Learning and Cybernetics, Qingdao, China, 11–14 July 2010; Volume 5, pp. 2454–2458. [CrossRef]
43. Coro, A.; Abasolo, M.; Aguirrebeitia, J.; de Lacalle, L.L. Inspection scheduling based on reliability updating of gas turbine welded structures. *Adv. Mech. Eng.* **2019**, *11*. [CrossRef]

Article

Methodology for Searching Representative Elements

Jure Murovec, Janez Kušar and Tomaž Berlec *

Faculty of Mechanical Engineering, University of Ljubljana, 1000 Ljubljana, Slovenia

* Correspondence: tomaz.berlec@fs.uni-lj.si

Received: 25 July 2019; Accepted: 20 August 2019; Published: 23 August 2019

Featured Application: The MIRE method was developed for a specific and practical case, where there is essentially minimal information about the company's material flow. It is suitable for smaller companies as it enables rapid analysis of a large quantity of mutually similar data. MIRE can relatively quickly give us a sufficient approximation of the material flow, which serves as the basis and the first step towards lean production, consequently increasing the company's competitiveness in the global market.

Abstract: Companies have to assure their share on the global market, meet customer demands and produce customer-tailored products. With time and production line updates, the layout becomes non-optimal and product diversity only increases this problem. To stay competitive, they need to increase their productivity and eliminate waste. Due to a variety of products consisting of similar components and variants thereof, a huge number of various elements are encountered in a production process, the material flow of which is hardly manageable. Although the elements differ from each other, their representative elements can be defined. This paper will illustrate a methodology for searching representative elements (MIRE), which is a combination of the known Pareto's analysis (also known as ABC analysis or 20/80 rule) and a calculation of a loading function, that can be based on any element feature. Results of using the MIRE methodology in a case from an industrial environment have shown that the analysis can be carried out within a very short time and this provides for permanent analysis, optimisation and, consequently, permanent improvement in the material flow through a production process. The methodology is most suitable for smaller companies as it enables rapid analysis, especially in cases when there is no pre-recorded material flow.

Keywords: representative elements; large data quantity; loading function; simulation; material flow optimisation

1. Introduction

Lean production as a production strategy is one of important paradigms of the 21st century which has been introduced in various variants practically in all companies in the world. A lean environment can minimise waste by elimination, merging or simplification of technologically unnecessary operations (stock, transport) and unnecessary elements of the work process (auxiliary and extra activities), that do not give any added value. Analyses [1] have shown that merely 20% of activities in a production process directly add added value, 20% of activities do not directly give any added value, but are needed in order to carry out activities that add added value, while the rest (60%) are the activities which do not add added value and are, therefore, indicated as waste.

If a technological operation is observed as a fundamental element of a lead time for the production of a component, it is determined that the lead time of an operation consists of: (1) transition time calculated as completion of the preceding operation until the beginning of performance of the observed operation; it represents as much as 90% of the time in the structure of the lead time of an operation, and (2) operation performance time which represents only a 10% share in the structure of the lead time [2].

Based on the above, the transition time is by far the most important in the structure of the lead time of an operation and it consists of a waiting time after a preceding operation, a transport time, and a waiting time before the observed operation [3].

Each production-based company has a certain arrangement of sectors and workplaces, which can vary in its optimal layout. A great potential in terms of reducing losses and consequently in raising competitiveness can be hidden in a sub-optimal layout. It is common that smaller companies add newly bought machinery to places where space was sufficient, despite the sub-optimal layout.

This finding calls for a need to reorganise the material flow within a production system in a way to have minimum [4] transition time between operations. This has a direct impact on the cost of a company's internal transport [5]. Francis, McGinnis and White [6] maintain, that 20% to 50% of direct costs can be attributed to the costs of material manipulation [7], wherein innovative organisational concepts have a decisive impact [8].

The material flow can be defined as a structured and organised movement of a material from point A to point B through a production system in compliance with a prescribed technological method under consideration of efficient use of space, savings in energy and human resources [5]. Material flow is a complex process of each production system and can be optimised without adequate tools for planning a material flow. Before optimising a material flow, the inventory must first be taken (the current situation assessed). A material flow usually consists of a huge number of elements (each with its own technological method) that are not identical but are similar in certain criteria. This is why families need to be formed on the basis of selection of representative elements. The quantity of data to be inventoried and analysed can thus be considerably reduced.

Bhosale and Pawar [9] have modified a mathematical model to optimise material flow optimisation of a flexible manufacturing system with a real coded genetic algorithm. In their article [10], the authors showed application of a statistical analysis and use of a group technology for the introduction of lean concepts into production systems in order to reach better leanness and consequently effectiveness of production systems. The group technology is also used for the analysis of a material flow in the process of converting a company having mass production and shop-floor configuration of workplaces into a cellular organisation of workplaces [11].

Due to a diversity of elements the conventional ABC analysis and the multi-criteria ABC analysis, such as [12,13], are no longer sufficient in our case to inventory the material flow.

Based on a literature review a new model called the methodology for searching representative elements (MIRE) was designed which assists us in processing huge quantities of data.

The result of the method is an approximation of the real situation, while deviations are checked by feedback. All five steps of the MIRE are repeated as necessary, until a satisfactory ratio between the approximation accuracy and time savings is obtained in the processing of data.

2. Methods

2.1. Literature Review

Group technology (GT) is an approach of production management, in which part families are formed from a plurality of various parts that are manufactured in a treated production system and that use the same group of machines to be transformed from an input material into the required final shape and property of the product.

The methods designed to search families of parts and the group of machines they belong to are treated in the literature as cell formation techniques. Based on a proposal of various authors Selim et al. [14] proposes the following overview of the cell formation technique:

(1) Descriptive procedures (part family identification, machine group identification, part families/machine grouping simultaneously);
(2) Cluster analysis;
(3) Graph partitioning;

(4) Artificial intelligence;
(5) Mathematical programming (linear programming, linear and quadratic integer programming, dynamic programming, goal programming).

Descriptive procedures part families/machine grouping were first represented by Burbidge as a Production flow analysis [15]. Dekleva et al. [16] extended the Burbridge's production flow analysis and called it the extended production flow analysis. The extended material flow analysis included two steps:

(1) Macroanalysis is analysing the material flow (flow of parts) on a macro level, i.e., between departments of a company. The goals of these analysis were: identification of extra paths between the departments of the company and their elimination (usually by changing a technological method) and determination of a standard material flow between the departments of the company.
(2) Microanalysis is included in a material flow analysis within an individual department. The goals of this analysis were: simultaneous identification of part families and groups of needed machines for their manufacturing and formation of production cells for the part families. A cluster analysis method and an incidence matrix method were proposed for the formation of cells.

Nouri et al. [17] propose, in particular, metaheuristic techniques such as: genetic algorithm, artificial neural networks, fuzzy set theory, colony systems, tabu search, branch and bound method, simulated annealing.

Nowadays, the goals of group technology and formation of various shapes of production systems (cells) need to be connected, especially with the goals of leanness, flexibility and agility of production. Due to market demands, a large number of different, yet mutually similar parts appear in the production process. That means that the production program contains a large number of different products that are similar to each other, as the companies emphasizes the possibility of making a tailored products for each customer to maintain its competitiveness in global markets. It is quite common in smaller companies that the record of material flow has never been made. They simply lack the time and the human resources, while trying to maintain acceptable production and make profit. With time and production line updates, the layout becomes increasingly non-optimal and the diversity of products only increases this problem.

So, it is not reasonable to analyse a material flow and a value flow on the entire amount of available data. Within thousands of pieces, representative parts (hereinafter representative elements) need to be selected. This is made possible by the methodology for searching representative elements, MIRE, which can relatively quickly give us a sufficient approximation of the material flow, which serves as the basis and the first step towards lean production, consequently increasing the company's competitiveness in the global market.

The methodology is most suitable for smaller companies that throughout their growth and development simply did not have the time, resources and competences to create a material flow inventory because their priorities were mainly focused on delivering the products and keeping up to date with the latest technological trends.

2.2. *Methodology for Searching Representative Elements (MIRE)*

Basic data for the implementation of the MIRE can differ, yet they usually refer to one selected influencing variable, e.g., mass, dimension, volume, size, material value, production time, added value. It is important that a selected variable is used throughout the methodology from its very beginning.

The MIRE methodology follows five steps represented in Figure 1. First, a time interval for data gathering and a creating criterion(s) are selected, which are then considered through the entire analysis process.

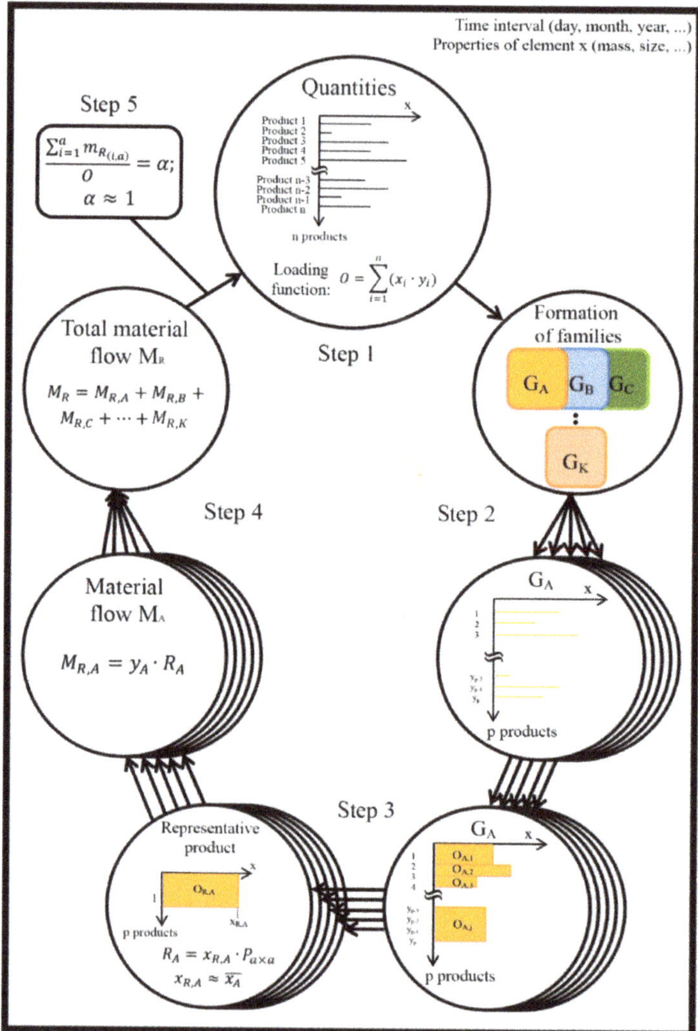

Figure 1. The methodology for searching representative elements (MIRE) circular flow diagram.

2.2.1. Step 1: Quantities

In a first step, an influential variable is selected that will serve as the basis for searching representative elements.

A production of n number of products is anticipated for the selected time interval, each of them having value x.

The loading function is defined as a product of the variable of one product with their own quantity and functions as a weight function (Figure 2). The total load in the selected time interval is:

$$O_I = \sum_{i=1}^{n} O_{I,i} = \sum_{i=1}^{n} (x_i \cdot y_i) \quad (1)$$

O_I—total load

n—number of products
x_i—value of the selected variable of the ith product
y_i—quantity of the ith product (always 1 before formation of families)

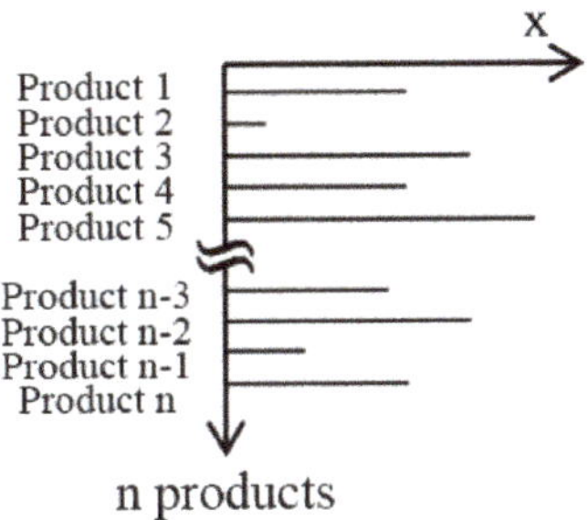

Figure 2. Total load.

2.2.2. Step 2: Formation of Families

The products having a similar sequence of technological operations are grouped into families (Figure 3), where the selected variable is not yet taken into consideration. The larger the number of families, the more precisely the material flow will be inventoried.

G_l—indication of family l
K—number of families; where: $K \leq n$
p—number of products in a family

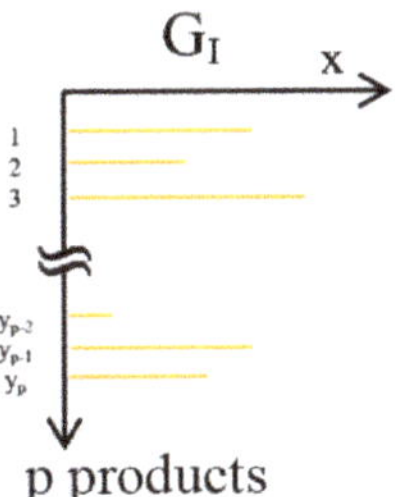

Figure 3. Formation of product families.

2.2.3. Step 3: Selection of a Representative Product of the Family

The family joins products with different values x, which have an identical or similar sequence of technological operations. The variables selected with an identical or similar value are grouped (Figure 4). The accuracy of grouping impacts the accuracy of material flow inventory. The more groups that are formed, the more accurate the inventory will be.

So, j groups are obtained and each group has a load $O_{l,i}$.

$$O_{l,i} = x_{l,i} \cdot y_{l,i} \tag{2}$$

$O_{l,i}$—load of the ith group of products having an identical or similar value of the selected variable within family l
$x_{l,i}$—value of the selected variable of a product in the ith group of family l
$y_{l,i}$—number of products in the ith group of family l

$$\overline{x_I} = \frac{O_I}{y_I} = \frac{\sum_{i=1}^{j} O_{I,i}}{\sum_{i=1}^{p} y_{I,i}} \tag{3}$$

$\overline{x_I}$—average value of the selected variable of a product of family l
O_l—total load of family l
y_l—number of all products of family l

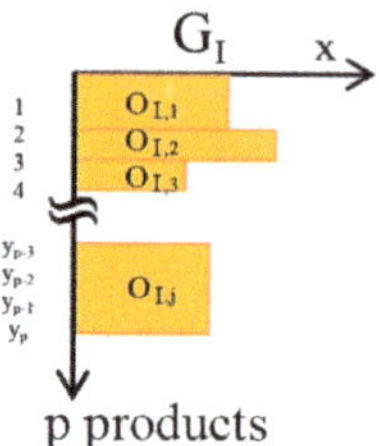

Figure 4. Grouping within a family.

The obtained $\overline{x_l}$ is used to search for such a product within family l, to which the value of variable x gets closest. In this way a representative product of family l, R_l, is selected.

The representative product should be assigned a path through the production and a sequence of treatment processes. This path should be written in a square FROM–TO matrix $\mathbf{P}_{a\times a}$, where a is the number of workplaces. Column or row a should represent the last workplace in the production (assembly, warehouse, takeover, etc.).

$$\mathbf{P}_{a\times a} = \begin{bmatrix} I_{11} & \cdots & I_{1a} \\ \vdots & \ddots & \vdots \\ I_{a1} & \cdots & I_{aa} \end{bmatrix} \tag{4}$$

The representative product of the Ith family thus has the value of the selected variable $x_{R,I}$ and path $\mathbf{P}$.

$$R_I = x_{R,I} \cdot \mathbf{P}; \quad x_{R,I} \approx \overline{x_I} \tag{5}$$

2.2.4. Step 4: Material Flow

The calculation of representative products of individual families allows us to define their material flow:

$$M_{R,I} = y_I \cdot R_I = y_I \cdot x_{R,I} \cdot \mathbf{P} \tag{6}$$

The total material flow of all families in this case is:

$$M_R = M_{R,A} + M_{R,B} + M_{R,C} + \cdots + M_{R,K} \tag{7}$$

2.2.5. Step 5: Checking Material Flow Approximation Accuracy

All representative products have been obtained and used to calculate an approximation of the total material flow. Since its accuracy is of interest, the equation below is used:

$$\frac{\sum_{i=1}^{a} m_{R(i,a)}}{O} = \alpha \tag{8}$$

$m_{R(i,a)}$—ith element of the ath column in matrix M_R.

The closest α comes to value 1, the more accurate the material flow M_R inventory is.

$$If\ \alpha = 1, \qquad then\ M_R = M \tag{9}$$

M—actual material flow.

The actual material flow or 100% accuracy of inventory of material flow M_R is obtained with the proviso that the number of families equals the number of all products in the selected time interval ($K = n$). In such a case, exactly one product is present within each family and this automatically means:

$$x_{R,I} = \overline{x_I} \quad and \quad j = 1 \tag{10}$$

As there are no deviations in the values of the selected variable and FROM–TO matrices of representative products, the material flow is inventoried with 100% accuracy.

The essence of this methodology to reduce the quantity of data to be processed in order to obtain a satisfactory material flow inventory. Therefore it is not reasonable, that $K = n$.

To obtain credible results consistency is needed:

- In classifying the products into families based on a criterion of sequence of treatment processes;
- In grouping the products within families based on the criterion (selected variable); and
- In selecting a representative product based on x_I.

As described, MIRE helps us process large amounts of data and it is especially useful for cases where the elements are different and similar at the same time, depending on the criteria (size, weight, cost, etc.). By creating families and selecting their representative elements, the amount of data, that needs to be processed is significantly reduced. In this case, due to the variety of elements, the classic ABC analysis is not sufficient [17]. The result of the process is an approximation of the real material flow, and the deviations are checked by a feedback loop. All five MIRE steps are repeated as needed, until a satisfactory ratio between the accuracy of the approximation and the saved time is achieved.

MIRE can also be used by companies that already had their material flow recorded, but need to quickly check how optimal their layout is. This is suitable after adding or updating the machinery or presenting a new product into the production line. In cases where the company never had their material flow recorded, MIRE can significantly shorten the time needed by reducing the quantity of data to be inventoried and analysed, because we do not need to consider all the material flow data, but only the elements, based on which we can satisfactorily describe the material flow. Needed input can be accumulated through positions of assembly drawings or any other source, that contains information about product parts, technological operations, etc.

3. Case Study

A Slovene company having 80 employees counts on adaptability and the possibility of manufacturing custom-made products, which provides an opportunity for being competitive on the global market.

Since the company does not have an Enterprise Resource Planning system (ERP system) or does rather not use it for production management, there is no material flow traceability. The company has a material flow inventory for positions of an assembly drawing; however, there is no material flow overview through the production to the assembly. So, there is no overview, through which processing phases blanks are transferred when leaving the main warehouse. Since a material flow inventory is needed to prepare a layout, this was a major problem. There was a huge quantity of data in the form of launching orders and structural product tallies. This is why a methodology had to be found, with which an accurate material flow inventory could be obtained in the simplest possible way.

An analysis of production assortments of companies in Slovenia has shown that data on masses of parts are the easiest to obtain; it was used in this paper as the main influencing variable x.

3.1. Application of the MIRE Methodology

The main advantage of the MIRE methodology is the reduction of the number of products, for which a path through the production needs to be traced in order to achieve a satisfactory inventory of the total material flow.

The period from 2012 to 2015 was selected for the analysis. Only the last year could have been included in the material flow inventory, which would reduce the quantity of data, however, a possibility of influence of potential fluctuation in orders would be increased. Inclusion of a longer period of time provides for a more precise annual sales average and more reliable anticipation of production load for the years to come.

The company produced approximately 3300 products in the past four years.

3.2. Selection of Representative Products

By using the clustering technology, the products were classified pursuant to their production mode into 96 groups. A two-dimensional ABC analysis was used [18] on criteria of product size and frequency of production. With the help of the employees, the products were divided into three main families and designated:

- Product (Table 1),
- Supplier product (Table 2),
- Module (Table 3).

The products were classified into 18 groups based on the quantity of produced products. Table 1 shows the data on the quantities produced in the period from 2012 to 2015.

Table 1. Quantities of the products produced in the period 2012–2015.

Name	2012	2013	2014	2015	Sum
Product 1	1	5	3	6	15
Product 2					0
Product 3	1				1
Product 4		3			3
Product 5	1	3	4	2	10
Product 6	3	6		1	10
Product 7	4	1	1	3	9
Product 8	1		1		2
Product 9	1	1	1		3
Product 10	3	2	5	1	11
Product 11		1			1
Product 12					0
Product 13	2				2
Product 14					0
Product 15			1		1
Product 16		1			1
Product 17					0
Product 18					0
Sum	17	23	16	13	69

For a further analysis, five families were selected from the entire family of products (18 products) The selection was based on experience and the quantities of products these families cover: Product 1, Product 5, Product 6, Product 7 and Product 10.

Suppliers supply a large quantity of various products (supplier products—77 families) for the analysed final product, yet many of them appear rarely. Table 2 shows the 10 most frequent products from suppliers, which are used for the analysis.

Table 2. Most frequent products from suppliers in the period 2012–2015.

Name	Quantity	Per Month	Per Year
Suppl. product 1	236	4.917	59.00
Suppl. product 2	86	1.792	21.50
Suppl. product 3	87	1.813	21.75
Suppl. product 4	85	1.771	21.25
Suppl. product 5	83	1.729	20.75
Suppl. product 6	37	0.771	9.25
Suppl. product 7	32	0.667	8.00
Suppl. product 8	31	0.646	7.75
Suppl. product 9	266	5.542	66.50
Suppl. product 10	30	0.625	7.50

Additionally, a module was selected that is integrated into all products (products from suppliers excluded). The modules differ only in size but are treated separately, because a buyer does sometimes not need it or orders it separately. As they are all practically only scaled versions of Module 1, weights were introduced on the basis of size/mass. A representative product of the module family was determined, which is shown in Table 3.

Table 3. Quantity and type of modules produced in the period 2012–2015.

Module Type	Quantity	Weight	Factor × Weight
Module 1	4	2	8
Module 2	2	4	8
Module 3	9	6	54
Module 4	5	8	40
Module 5	3	12	36
Module 6	3	16	48
Module 7	5	20	100
Module 8	11	24	264
Module 9	9	32	288
Module 10	2	40	80
Module 11	1	64	64
Sum	54		990
Average module weight		18.33333	

As a representative product, only one module is selected, the weight of which comes closest to the average weight of the module. In this case, Module 7 with weight 20 came closest to the average weight of the module which amounts to 18.3. A weight function method was used to determine all representative products of the selected families.

4. Simulation

A production line with 356 different product groups was simulated, creating more than 44,000 individual products. The properties of each were: produced quantity, mass, the sequence of technological operations, size, cost and production time. This represented a common year of production in a smaller company. Data was written in a spreadsheet, then read and handled with a code written in MATLAB (Matlab 2018a, The MathWorks Inc., Natick, MA, USA). The presented simulation shows how we can choose a small fraction of products according to desired variables and still achieve a sufficient portion of the whole material flow.

A simulation by all steps of the MIRE methodology for the selected production will be shown below. In a first step, an influencing variable needs to be selected, which is a product mass in our case. Figure 5 graphically illustrates all products of the production as a function of mass.

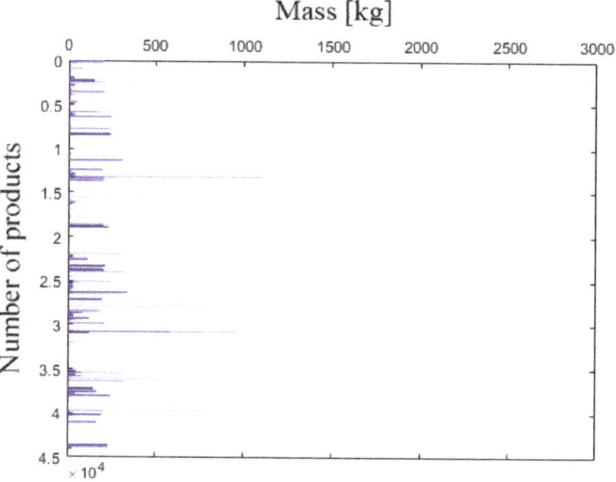

Figure 5. Simulation of all products.

Figure 5 shows an entire population of products of the selected production at the selected influencing variable (mass). The arrangement in Figure 5 is not ordered by size, it is random. There is huge dissemination of a number of products based on the criterion of product mass since the mass ranges from several kilos to more than two tons. The illustrated case represents a typical production assortment of a company that produces custom-made products.

This step is followed by grouping of products within a family (Figure 6) and selection of a representative of the family that will make the best inventory of the products within a family (Figure 7) as the representative of this family.

Figure 6 illustrates a distribution of products by mass within a certain family that was formed based on the sequence of treatment processes. More than 500 products are classified in 25 groups.

If a closer look is taken at each group, an average mass of one product of the entire family can be determined (Figure 7).

The representatives of groups within a family and their mass are shown in Figure 7. The red line represents an average mass of a product in a family. A representative of group 13 comes closest to this value and this is why it is selected as the representative of the family.

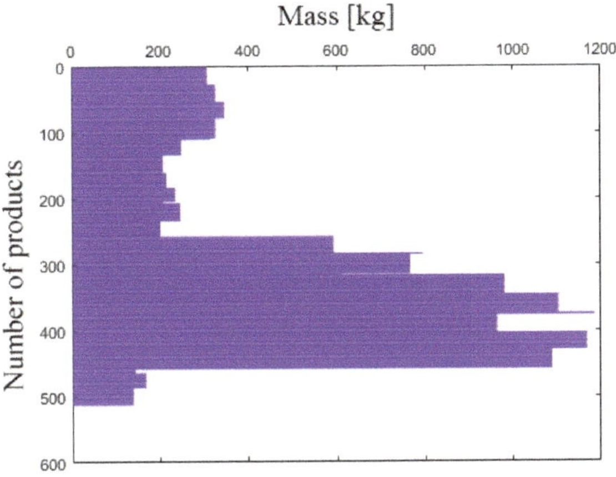

Figure 6. Simulation of grouping of products within a family.

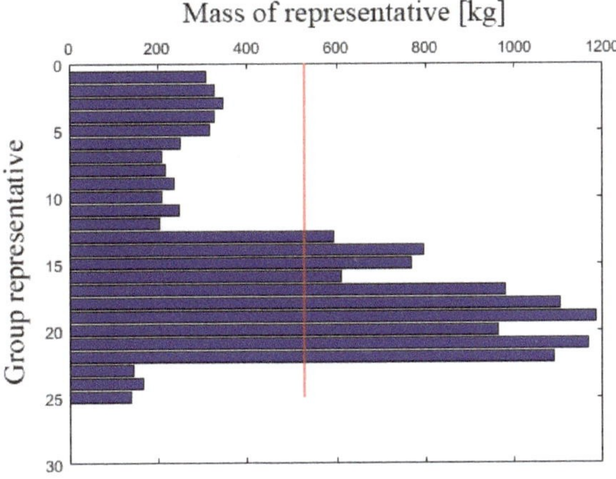

Figure 7. Simulation of selection of a representative product within a family.

Then follows simulation of descending distribution of representatives by share under consideration of the criterion of the mass (Figure 8).

The graph in Figure 8 can be used to determine how many representatives will be selected and their material flow inventoried to include a satisfactory share of the entire production assortment. The criteria of size, price and production time are also shown.

An increase in the share of production time and size is noticed with the twelfth heaviest representative, and it will therefore be the most reasonable to select the 12 heaviest representatives and define their material flow. In this way, more than 80% share of the mass of all products, almost 60% share of the size of all products, approximately 50% share of the total production time, and approximately 45% share of the price of all products would be captured.

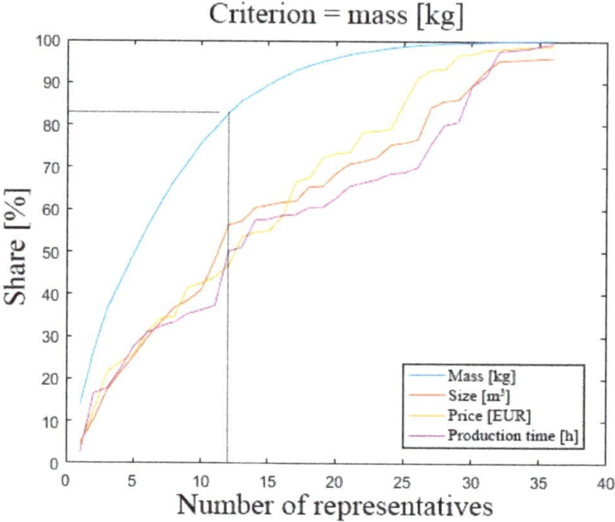

Figure 8. Simulation of descending distribution of representatives by several criteria.

Figure 9 shows a graph of selection of a number of representatives in a multi-criteria selection. The important variables are the mass and size of a product in a ratio of 3:2, which serves as an example of approximation of criteria importance regarding transport and stock. As evident from the figure, the reasonable number of representatives to be selected could be 6, 11 or 17, as the processing time/share of the inventory ratio is the most beneficial. This depends on the share of the inventory of all variables, with which we are satisfied. The larger the number of selected representatives, the higher share of the production assortment will be included. If the first 11 most important representatives are taken as an example based on the mass and size (in a 3:2 ratio), an approximately 80% share of the total mass of all products, an almost 60% share of the size and some 30% share of the price and production time of all products are inventoried.

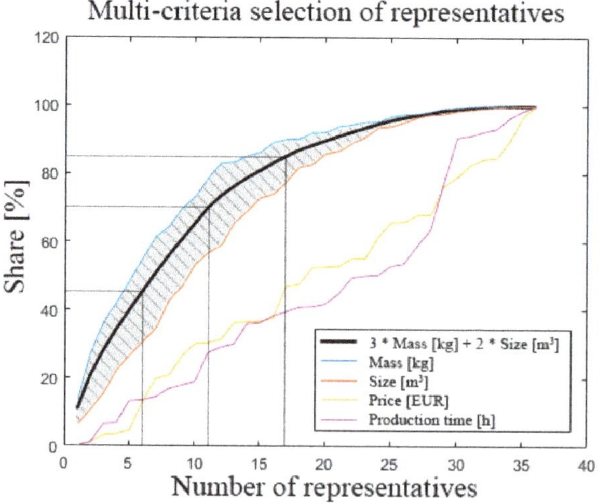

Figure 9. Simulation of a descending distribution of representatives based on two criteria with different weights.

The black line represents an order of the representatives which are distributed from the most important to the least important one based on both criteria: mass and size in a 3:2 ratio. If these two criteria would be equally important, the black curve would be a line of symmetry of the red and blue lines or a centroidal axis of the hatched area, but we see that it comes closer to the curve of the mass.

5. Results

The methodology for searching representative elements provides a satisfactory material flow inventory. In fact, it proves to be a very efficient way for the processing of a large quantity of mutually similar data. The simulation shows how MIRE manipulates product information through each step. The generated data represents a typical production assortment of a company that produces custom-made products. Figures 8 and 9 describe how the share of selected variable increases together with the number of representatives of each product family. The increase of the share of other variables can be seen, too, and an example of one and two criteria simulation is illustrated.

Out case study did not include any pre-recorded material flow, thus only an approximation of the material flow share could be done with MIRE. Sufficient data had to be obtained manually from assembly drawings and it would be very time consuming to gather all the information needed to calculate the entire material flow. A solution was needed quickly, which led to the creation of MIRE methodology. Table 4 shows that the representative products inventory almost a half of the total material flow. There may be certain deviation, yet it can be neglected if we bear in mind that only 0.5% of all products were selected for the 44% share of the inventory. The work is reduced by almost a factor of 100, while the obtained material flow is more than sufficient to design a new layout because it includes the most frequent, the largest and the most important products of the company.

Table 4. The share of material flow inventory and the share of the number of products.

Period 2012–2015	Number of Products	Number of Families	Material Flow Share	Time Consumed
All products	3300	96	100%	estimation: 4–6 months
Representative products of selected families	16	16	44%	2 weeks
Share	0.5%	17%	44%	8%–16%

MIRE is estimated to speed up the process of creating the material flow inventory up to 10 times. It can be also be used to check how optimal is the current layout. Instead of weeks, the material flow can be calculated in hours, saving time and consequently money. Although the output does not cover 100% of the material flow, the final share of it is sufficient as we can choose and adjust and multiple variables and its weights until they meet our needs and achieve high enough certainty of the result.

6. Conclusions

There are many efficient methods designed to search families of parts. They vary from relatively simple to very sophisticated, but they all have one thing in common: a large amount of data. In our case, the company did not have a pre-recorded material flow. There was no ERP system, so we had to be creative in our search for relevant product information. We gathered the necessary data from assembly drawings manually. With MIRE methodology we were able to find the sufficient data for the calculation of an approximation of the total material flow.

The MIRE saves plenty of time in processing a huge quantity of data. It is mostly applicable for cases, where the elements differ from each other based on some criteria and are similar to each other based on others. It can be of great help to the companies that do not have a digital observer of inventory of material flow in the ERP system, because it can give a sufficient approximation of the actual situation. The MIRE result helped in conceiving a suboptimal production layout that will serve as a basis for re-organisation of the production of the company.

Having proved that the MIRE is applicable and efficient, it is reasonable to develop it further. The steps, in which weight functions are used, can be even more precisely defined.

Simulations would help us detect the influence of deviation of a representative product from an average product, the influence of the number of families which is inversely proportional to the number of elements in an individual family, and the influence of accuracy of grouping of products within a family. Input data, such as number of workplaces, number of families, and number of all products would additionally be changed and their correlation with other factors could be monitored.

Ideally, a mathematical model could create graphs and tables that will be easy to use and from which it would be easy to deduct what share and reliability of a material flow inventory can be anticipated based on known input parameters. Simultaneously with simulations, the MIRE would be used on the data in the companies which have an accurate material flow inventory as an upgrade to the ERP system. The simulated and real results could be compared for an even more thorough understanding of MIRE's advantages. This could be done in a company that has a complete overview and tracking of its material flow. In that case, many other methods have already been developed, but all of them require an organised database. The MIRE method was developed for a specific and practical case, where there is essentially minimal information of the company's material flow. Therefore, comparison to other methods is a little difficult and unwarranted. Regarding further development of MIRE methodology, some programming knowledge could convert the final product into computer software, into which a person could enter input parameters, select the desired material flow, and share its accuracy. This would result in instructions that would suggest which families should be selected and what should the representative products be, to which the FROM–TO matrix would need to be assigned.

Author Contributions: J.M. and T.B. conceived and designed the experiments; J.M. performed the experiments; J.M., T.B. and J.K. analyzed the data; J.K. contributed analysis tools; J.M. and T.B. wrote the paper. All authors have read and approved this paper for submission.

Funding: This research received no external funding.

Conflicts of Interest: The authors declare no conflict of interest.

References

1. Ward, A.; Runcie, E.; Morris, L. Embedding Innovation: Design Thinking for Small Enterprises. *J. Bus. Strategy* **2009**, *30*, 78–84. [CrossRef]
2. Wiendahl, H.P.; Reichardt, J.; Nyhuis, P. *Handbool Factory Planning and Design*; Springer: Berlin/Heidelberg, Germany, 2015.
3. Zupan, H.; Herakovič, N.; Starbek, M.; Kušar, J. Hybrid Algorithm Based on Priority Rules for Simulation of Workshop Production. *Int. J. Simul. Model.* **2016**, *1*, 29–41. [CrossRef]
4. Alvarado-Iniesta, A.; Garcia-Alcaraz, J.L.; Rodriguez-Borbon, M.I.; Maldonado, A. Optimisation of the Material Flow in a Manufacturing Plant by Use of Artificial Bee Colony Algorithm. *Expert Syst. Appl.* **2013**, *40*, 4785–4790. [CrossRef]
5. Mayers, F.E.; Stewart, J.R. *Motion and Time Study for Lean Manufacturing*, 3rd ed.; Prentice Hall: Engelewood Cliffs, NJ, USA, 2001.
6. Francis, R.L.; McGinnis, L.F.; White, J.A. *Facility Layout and Layout: An Analytical Approach*; Prentice Hall: Engelewood Cliffs, NJ, USA, 1992.
7. Singh, S.P.; Sharma, R.R.K. A Review of Different Approaches to the Facility Layout Problems. *Int. J. Adv. Manuf. Technol.* **2006**, *30*, 425–433. [CrossRef]
8. Koren, R.; Prester, J.; Buchmeister, B.; Palčič, I. Do Organisational Innovations Have Impact on Launching New Products on the Market? *Strojniški Vestnik J. Mech. Eng.* **2016**, *62*, 389–397. [CrossRef]
9. Bhosale, K.C.; Pawar, P.J. Material Flow Optimisation of Flexible Manufacturing System Using Real Coded Genetic Algorithm (RCGA). *Mater. Today Proc.* **2018**, *5*, 7160–7167. [CrossRef]

10. Božičković, R.; Radošević, M.; Ćosić, I.; Soković, M.; Rikalović, A. Integration of Simulation and Lean Tools in Effective Production Systems—Case Study. *J. Mech. Eng.* **2012**, *58*, 642–652. [CrossRef]
11. Suzić, N.; Stevanov, B.; Ćosić, I.; Anišić, Z.; Sremčev, N. Customizing Products through Application of Group Technology. *J. Mech. Eng.* **2012**, *58*, 724–731. [CrossRef]
12. Douissa, M.R.; Jabeur, K. A New Model for Multi-Criteria ABC Inventory Classification: PROAFTN Method. *Procedia Comput. Sci.* **2016**, *96*, 550–559. [CrossRef]
13. Liu, J.; Liao, X.; Zhao, W.; Yang, N. A Classification Approach Based on the Outranking Model for Multiple Criteria ABC Analysis. *Omega* **2016**, *61*, 19–34. [CrossRef]
14. Selim, H.M.; Askin, R.G.; Vakharia, A.J. Cell Formation in Group Technology: Review, Evaluation and Directions for Future Research. *Comput. Ind. Eng.* **1998**, *34*, 3–20. [CrossRef]
15. Burbidge, J.L. Production Flow Analysis. *Prod. Eng.* **1971**, *50*, 139–152. [CrossRef]
16. Dekleva, J.; Kušar, J.; Menart, D.; Starbek, M.; Zavadlav, E. Extended Production Flow Analysis. *Robot. Comput. Integr. Manuf.* **1988**, *4*, 63–68. [CrossRef]
17. Nouri, H.; Tang, S.H.; Hang Tuah, B.T.; Ariffin, M.K.A.; Samin, R. Metaheuristic Techniques on Cell Formation in Cellular Manufacturing System. *J. Autom. Control. Eng.* **2013**, *1*, 49–54. [CrossRef]
18. Starbek, M.; Petrišič, J.; Kušar, J. Extended ABC analysis. *Strojarstvo J. Theory Appl. Mech. Eng.* **2000**, *42*, 103–108.

MDPI
St. Alban-Anlage 66
4052 Basel
Switzerland
Tel. +41 61 683 77 34
Fax +41 61 302 89 18
www.mdpi.com

Applied Sciences Editorial Office
E-mail: applsci@mdpi.com
www.mdpi.com/journal/applsci

www.ingramcontent.com/pod-product-compliance
Lightning Source LLC
LaVergne TN
LVHW070614170726
843515LV00004B/73

* 9 7 8 3 0 3 9 4 3 9 6 1 4 *